THEY LIVE IN THE SKY

Trevor James Constable

(originally written under the pseudonym Trevor James)

THE BOOK TREE
San Diego, California

© 1958
Trevor James Constable
All rights reserved

ISBN 978-1-58509-576-6

Cover layout
Mike Sparrow

Cover photo copyright
by Pindyurin Vasily

Published by
The Book Tree
P.O. Box 16476
San Diego, CA 92176
www.thebooktree.com

We provide fascinating and educational products to help awaken the public to new ideas and
information that would not be available otherwise.
Call 1 (800) 700-8733 for our *FREE BOOK TREE CATALOG.*

To My Mother

ACKNOWLEDGEMENTS

The author wishes to extend his thanks to the following persons, whose assistance, advice and suggestions have been invaluable to the preparation of this book:

Dr. Franklin Thomas, for technical advice on optical and occult subjects.

James O. Woods, for indispensable personal assistance with both photography and finance.

Manon Darlaine, for research aid and advice.

J. K. Allen, for helpful spiritual suggestions.

Frank Der Yuen, for the loan of equipment and for much encouragement.

Bob Bucher, of Paxton Camera Shop, North Hollywood, for advice offered in the face of frustrating non-conformity on our part.

Ernst Lehrs, Ph.D., whom I have never met, but whose book, "Man or Matter" has given indispensable guidance.

And to all those I have never seen, who have helped.

CONTENTS

Preface	13
Chapter I. Communication Is Possible	17
Chapter II. "Friends"	31
Chapter III. "Foes"	51
Chapter IV. "Green Men, Monsters and Skeptics"	83
Chapter V. Science, the Infallible Phalanx	109
Chapter VI. Are there Assaults from the Invisible	133
Chapter VII. "Don't Call Us, We'll Call You"	159
Chapter VIII. Adventures in the Desert	179
Chapter IX. Frankenstein in Blue	192
Chapter X. The Photographs	214
Chapter XI. Spacemen, Vibrations and Levity	241
Chapter XII. Retrospection and Speculation	255
Bibliography	271
Appendix containing Affidavits	273

PUBLISHERS NOTE

For centuries Unidentified Flying Objects have flashed across our skies and sometimes have hovered or floated about, but seldom close enough for accurate observation. Usually there have been no witnesses to a close encounter or any objective evidence. Likewise countless reports of messages have been made available, but without any proof of the identity of the intelligence communicating or the truth of the message itself.

Until we saw the manuscript and pictures of "They Live in the Sky" we were of the opinion that there was little to be added to the abundant literature dealing with Unidentified Flying Objects, unless a Flying Saucer were to land on the White House lawn.

However we saw at once that this book presented a new and original approach to the solution of the mystery surrounding this subject. There is something refreshingly new about honest and unbiased research and reason applied to such a mysterious and controversial subject.

The author of "They Live in the Sky" set out to discover the truth for himself and applied modern photochemical methods to obtain some measure of objective recording of his research. While the results are not all that could be desired, they are amazing and convincing. This pioneer work has been done entirely without outside assistance, and when it came to a choice between a home and a new car, these desires were sacrificed and the money spent on further research. The author found one faithful friend to accompany him on the long desert trips and assist in the photography involved. We feel the public owes them a great debt of gratitude for making available the results of their lengthy and arduous research.

The reader will find "They Live in the Sky" a very entertaining narrative, of the author's personal experiences, together with a very enlightening and convincing exposition of the many phases of the UFO mystery. Nothing like it has ever been offered from any other source.

—Franklin Thomas

PREFACE

It is virtually impossible to keep this type of book abreast of current events in the UFO field. It has also been impossible to keep it abreast of our own researches, especially in the field of photography. This book should therefore be looked upon solely as a step along the way, and in no sense the final word from this source on these matters. Precipitate criticism, especially from scientific quarters, may well result in crushing rejoinders, evidentially backed. The work has not stopped and will not stop, nor will the discoveries.

The art of photographing the UFO requires a book by itself. Only an outline of the system used by us until the end of 1957 is covered in Chapter X. A second book, dealing specifically with this subject and tentatively titled "How to Photograph the UFO" is now in preparation. Its publication should enable many thousands of persons all over the world to photograph the UFO in a gigantic do-it-yourself project. Its results should give all air forces something to think about.

In the twelve years since modern Ufology began, certain basically erroneous concepts have grown into acceptance by devotees of the phenomena. Re-evaluation is needed in many fields. Blanket terms, such as "the space people" need to be eliminated, and careful attention paid not only to the invisible animals that fly in our air. but to the various orders of beings who either visit us or who have their natural habitat in invisible domains surrounding and interpenetrating our own.

It is also childish to assert, and incorporate into the foundations of ufology, the automatic conclusion that these UFO *must be* interplanetary. Not only has our own earth not been scientifically eliminated as a source, whether that source be on or *below* the surface, but our own atmosphere has not been examined properly. The creatures I have photographed clearly point the direction that must be followed in investigation of the UFO before we permit the interplanetary answer to be crammed down our throats by its adherents.

The investigator who is unalterably convinced, as I am, that man is a spiritual organism, is always confronted with the problem of proof. The onlooker consciousness, which holds the universe to be mechanical, has enthroned the eye over the kingdom of reality.

Nevertheless, various materials and devices produced by conventional science, used in unorthodox ways can provide externalized proofs in our day and age of things known to mystics and esoteric philosophers for scores of decades. The photograph of the "Amoeba" is proof of this. All that this book contains relative to the UFO has been obtained by two men, relatively uneducated but spiritually awakened, on less than one month's salary for a movie star or a scientific gentleman presiding over an empire of destruction.

There are those who will say that I have lied, or that I have faked. Governmen investigation, which is welcomed, will reveal that I have been truthful and honest, and the passage of time with ensuing discoveries, will reveal me as accurate.

There is a powerful group in the United States dedicated to keeping UFO investigation scientific and respectable. This group wants to "make a lady" out of what has become a somewhat shabby subject. The group disavows all contact stories and equates retired military or naval rank with ability in original thought. The spiritual side of the UFO is taken care of by inducing a minister or two with a high-class group, by compromising with orthodoxy has gone congregation to join in its pronouncements. This precisely nowhere since its inception, despite its "respectibility." The answers to the UFO are to be found in casting off dogma, not in dressing it up in a new suit.

On the matter of space animals, orthodox astrophysics has uttered its croak through Dr. Donald Menzel, Professor Emeritus of Astrophysics at Harvard University: "Astrophysics knows nothing of such things." But they exist! So much for orthodox thinking.

In the New Age dawning, the spiritual scientist will come into his own because movement toward the Godhead is the direction of human progress. The spiritual scientist is in tune with this movement, and sets himself to aiding it in whatever manner suits his talents.

It is my hope that my work will not only start new investigations along spiritual lines, but that it will bring trained scientific intelligences into the conscious service of spiritual science. By this means, man's final emancipation will be hastened, and to this end, my work is dedicated.

"For we wrestle not against flesh and blood, but against principalities, against powers, against the rulers of the darkness of this world, against spiritual wickedness in high places."
— St. Paul's Epistle to the Ephesians, Chapter 6:12

THEY LIVE IN THE SKY

CHAPTER ONE

COMMUNICATION IS POSSIBLE

"Seek and ye shall find" — Jesus.

It is unlikely that the truth about flying saucers will reach the public through governmental agencies. The reason for this is not to be found in any malicious desire to conceal the truth, but simply in the fact that neither the governments nor any individuals know the truth about the U.F.O.

This book is positively *not* the truth, the whole truth and nothing but the truth about the UFO, but it does bring forward to the public consideration *certain truths* about the UFO that will help build a larger mosaic. The truths about certain of the UFO herein revealed were obtained the hard way, and involved considerable expenditure of time and money as well as some risk.

The starting point of all that this book contains is that the truth belongs to those who seek it, on the wise principle of "seek and ye shall find." Seeking is not for the timid, since for each thing that is found, one can count on running down many blind alleys first.

Early in my own UFO studies, I concluded that officialdom throughout the world was hopelessly at sea over the whole UFO phenomenon, and that its evidential aspects were so completely beyond conventional science that an ordinary man like myself might have as good a chance of penetrating the mystery as anyone else.

Existing theories about life, matter, physical laws, and propulsion are likely to prove nothing more than a very rough guide in getting at the truth of the UFO, for reasons which we shall presently discuss.

A lack of formal scientific training is nowhere near as severe a handicap as purblind devotion to existing theories, when investigation of the UFO is being undertaken by an individual or a group. This is my opinion, formed after many months of investigation.

The latter day advent of the UFO was regarded by me as being definite proof of my lifelong belief that intelligent life does exist elsewhere in the universe, a belief which astrophysicists, astronomers and even ordinary medical doctors increasingly share with esoteric philosophers, who have never believed anything else. For generations, however, astronomical theory had held life could not exist here, or there, or elsewhere in the Universe because certain conditions had not been fulfilled. These mechanistic concepts have pretty well gone by the board in recent years, for there is no other valid explanation for the presence of the UFO, than that there is intelligent life elsewhere in the universe. One at a time, and usually after contact with UFO data, astronomers are getting on the "life on other worlds" bandwagon, leaving only a few diehards to pound away with the calcified theories of yesteryear.

Astronomy has been seeking to get off the hook because it is an observational science, and the observational evidence now available of intelligent life elsewhere is so overpowering that the old theories are untenable. The activity on the moon, for example, indicating physical structures being erected, finally puts the hammer on the honored theory that the moon is a dead ball of matter in the heavens.

Thus observation, the evidential flail the astronomer has laid on the theories of others in the past, is now being used by intelligent persons to beat the life out of some of the most dearly beloved theories in science.

Flying saucers, the UFO, by observation both electronic and visual are "fait accompli" and arguments as to whether they do exist or not are idle, time wasting and an obstacle to further investigation. In this book, we presume that the UFO are real, and get down to the discussion of the queries "Where do they come from?" "What are they?" and "What do they want?" And the less actively pursued question "What is the *nature* of their reality?"

Like almost everyone else, my own interest followed routine channels, which included the books of Keyhoe, Cramp, Wilkins, Jessup, Adamski and others. Again like almost everyone else, I found the variation of viewpoints to be very wide, and the confusion tremendous in these standard books on the subject. The interpretations placed upon the observed phenomena also differed widely.

Each of the saucer writers seemed to have pet personal theories regarding the UFO, into which he sought to press the facts. In a couple of years of study, I frankly did not know whether I was coming or going. Were the UFO friendly or hostile? Were they manned by humans like us or a different type of being? These questions and others still puzzle almost every UFO student and enthusiast, and most of them remain unanswered.

The point of departure for me with standard UFO writings came over the question of communication with these UFO and contact with the occupants of the craft, if indeed they were craft at all.

The reason I believed that communication must be possible with the craft was that in rather extensive travels around the world I had noted that civilization in every country was directly proportional to communications development. The reason I had paid attention to communications development was that I was at the time of my journeyings a radio officer in the British Merchant Navy.

In parts of the Orient and Middle East, a telephone is still a very rare instrument. If you can find one, a call to a point a hundred miles away may take a day or two, or you might never get it. Telegraph communications are scanty, and radio almost non-existent.

One reaches bottom where communication is concerned in those portions of Africa where the tree trunk drum or the human runner still serve as the means. In these locales, material civilization is at its most retarded.

This fact of communications as the index of civilization is not widely appreciated in the United States, where through habit, we speak, write, and project the human image over thousands of miles with a familiarity bound to breed contempt for it. Far from being happy with our communications, including radio and television in their exquisite refinements, manufacturers are driven to improve and expand their services by relentless competitive pressure for improved ability to communicate.

The ability to communicate on a world-wide scale in a variety of media is available to every American, and in this country we have the highest *material* civilization known on this planet.

It is obvious from their observed speeds and manoeuvres, their advanced propulsion systems as now acknowledged, their ability to appear and disappear, that the UFO are undeniably products of a civilization immeasurably advanced beyond our own, *technologically speaking.* While we boast in our newspapers of airplanes capable of fifteen hundred miles per hour, other craft, perhaps not of this world, have been electronically measured travelling six times this speed, and more, in our atmosphere.

In my view, it is irrational to admit the existence of such advanced vehicles and then deny to their occupants the ability to communicate on a scale far beyond anything we now have. If communication is the handmaiden of technological progress, as is the case in our own earthly civilization, it is logical to assume that it will be true of other worlds.

We must expect that the intelligences occupying or directing these strange craft will have mastery of all methods of communication, including all that we now have on earth, and perhaps many systems that we could not comprehend.

They may look upon our radio as humorously primitive, and regard our best radar and electronic detection apparatus as infantile. We must also not forget that they may have developed *natural* communication methods to a high degree, since all that we have in the way of communications is nothing more than the externalization of the contents of our minds. The radio transmitter was conceived in the mind of man, then externalized, that is, fabricated. In a sense, the radio transmitter is nothing more than materialized thought.

Communication engineers who vault upon their high horses when these concepts are advanced would do well to review the very short history of electronic communication on this planet. There are living humans today who were grown men and women before radio in any form was known. Let that reminder suffice for expanding egos.

My own viewpoint at the time I began my investigations was that the probability was very great that these UFO beings would have communication methods as far beyond ours as those of modern America are beyond the tree trunk drum. I have found nothing subsequently to nullify this viewpoint, and overwhelming evidence that it is the case.

In the light of the above reasoning, I found it absurd that so many otherwise able and intelligent men writing on the UFO could be so adamant in saying that there had been *no communication with saucer beings*. On the one hand, as ufologists and saucer proselytizers, they recognized the presence in the atmosphere of the products of an advanced civilization, while on the other they virtually denied the controlling intelligences any ability in communication. The most unfortunate part of this viewpoint is that it was and is held today without investigation. Investigation will reveal the theory to be false.

For myself, I sought to "become as a little child," by sweeping out of my mind accumulations of dogma and prejudice. I began by placing myself in the position of the UFO beings, and wondered what my course of action would be if I wanted to contact humans on the surface of the earth.

Could I, for example, go straight to the President of the United States? I felt that this would not be feasible, for the moment the president met with his advisers and told them a "space being" had communicated with him, the conclusion would doubtless be drawn that the pressure of the job had unhinged his mind. The same applies to all top government officials.

Science, in the orthodox sense, would also have to be counted out quickly. All that dogma, all those misconceptions, all those brilliant minds banged, barred and bolted against anything radical. The UFO being might reason that this type of mind, which has been the major obstacle to innovation since the time of Galileo is the poorest possible soil for new cosmic ideas.

It seemed to me that the UFO beings would finally be compelled to start at the bottom and work up, rather than at the top and work down. Working from the top down is the system favored by some people, but I considered that the UFO beings would have *no choice* but to get down to the grass roots and slowly let their presence, purposes and ability to communicate seep through the general population. By this means, governments would finally be forced to take cognizance of the claims, especially if the lower orders of us human beings were given information that would eventually be proven true.

It was in this way that I came to visit Giant Rock, California, a desert airstrip in an isolated spot between Victorville and Twenty-nine Palms. With a long history as a landing ground for aliens in pre-war days and as a burial ground for Indians in pre-paleface days, Giant Rock is difficult of access over dirt roads that try the springs of the stoutest automobile. A dry lake bed and a shingle airstrip run uphill to the foot of the Giant Rock, a seventy foot high boulder with enormous girth. Behind the rock is a four hundred foot high pile of smooth, round boulders.

In this arid, flinty locale there dwells a man who seems to harmonize with his environment. Middle aged, with sandy, thinning hair, George Van Tassel is one of the pioneers of communication with the UFO.

Communication? No radio towers or antennae or electronic equipment clutter the area, for communication is carried on by telepathy, or thought transference, this skill, known to and used by the ancients and basically possessed by every human being, is one of the God-given faculties. It has atrophied in most of us through centuries of non-use.

Beneath the Giant Rock, a large chamber has been hewn from the ground and furnished with chairs and tables. From this room, Mr. Van Tassel carries out his communications with the saucer beings.

In complete darkness in this cavern under the Giant Rock, and after preparatory measures closely resembling those of a seance, there booms forth from Van Tassel's direction a voice that is most definitely not his own. "I am Hulda, your people will soon witness more fireballs, which we are dropping as nullifiers. Greetings of love and peace to you . . . "

More messages follow, in voices which vary in accent, in timbre, in a manner that would be beyond the ability of even the most talented actor. These discourses deal with a variety of subjects, including life on other planets, UFO propulsion and always with atomic power.

In my view, the entities speaking through Mr. Van Tassel were of a standard of intelligence far beyond his own. This is not intended to be derogatory to Mr. Van Tassel, who, while a likeable personality, is not a highly educated man and does not claim to be such. These entities speak with a grammar, and exhibit sentence construction and a vocabulary far beyond Mr. Van Tassel's attainments in these fields.

The use of Mr. Van Tassel's personal physical facilities for these beings to manifest indicates that they are able to manifest here on earth only through the agency of suitable humans, although this is not yet entirely proven.

Mr. Van Tassel is willing enough to give an earnest seeker the method of preparing oneself for this type of contact. He told me when I asked him, and made no claims for patent rights or special talent. He did warn of possible ill effects, but fired with enthusiasm, I paid his cautioning little heed.

In the light of my own later experiences with telepathy, I do not consider it wise to dispense the information regarding preparation for it to all and sundry. It can be dangerous, and there are those who will dabble in these things who are totally unprepared for such activity. I know, because I was just that type of person myself. Having asked for what I got, I have no complaints, but I will not myself be responsible for dispensing the same information to others, and hence exclude it from this text.

Mr. Van Tassel has no "racket" or gimmick at the Giant Rock. He has nothing to sell, and together with his family and associates, appears dedicated to further work in the UFO field. The idea that he is cleaning up financially with his tiny cafe, often broached by UFO writers who have never been there, is fatuous.

I would like it understood that while I am more than glad to pay appropriate tribute to George Van Tassel for his part in my early experiences, I do not necessarily agree with him today regarding the UFO nor do I support his projected political ventures.

In the months that followed I sought contact consistently. As an investigator, I felt this was the only fair way to test the suggestions Mr. Van Tassel had given me. Until I had, I was in no position to pass judgment on this contact business. Like all applications of spiritual science, *investigation* inevitably involves *participation*. One cannot afford the luxuries of the onlooker consciousness, nor stand the penalties its misconceptions inflict upon spiritual growth. One participates, one experiences, and then one *knows*. The truth is within the one who has had the experience, and it may not be contradicted or negated, not by high domed pontiff, by atheistic scientist, nor by military officers active or retired. There is only one way to test the validity of UFO communication, and that is to try it. He who has not tried it is not qualified to render judgments on its possibilities or probabilities.

The months slipped past, without any results of any kind. However, I resolved that I would persist for a minimum period of six months, discouraged or not. Any other attitude would have been unfair and inconclusive.

I continued to study the UFO of course, seeking to find the central theme of coherence that would make the pieces fit. One night at about ten o'clock I was reading Keyhoe's "Flying Saucer Conspiracy" with some care to details of one particular chapter. Suddenly, I was seized by an overpowering impulse, which I now realize was a tremendous thought impression, to pick up a pen and write. Obeying the impulse, I picked up a pencil and began writing on the cover of a "paper back" book, the only thing handy on which I could make a mark. My arm was impelled and controlled by an unseen force, and I wrote a scrawl which was largely unintelligible across the back of the book. At the bottom of the cover, I distinctly wrote the word "discontinue", at which time the force on my arm was released just as though a switch had been thrown.

I discussed this happening with my wife, who was somewhat startled. I stood in the middle of the floor gesticulating and giving vent to my amazement, when a similar force was exerted upon my head, drawing it upwards and backwards with surprising power.

By a great effort of will, I was able to return my head to the normal position. The moment I ceased willing it there, the force drew it upwards and backwards once more. Suddenly, as before, the force cut off as though switched, and my puzzlement was now doubled. I discussed it briefly with my wife, and a few minutes later had an overwhelming desire to go outside. Not wishing to further alarm my wife, I made the excuse that I had to go outside and move the car.

As I moved to the door, my right hand rose to the knob by this self-same force, and closed around it. Once again, I could *will* it to my side. I turned my body slightly so that the left hand was nearer to the knob, and it too rose and grasped the knob without any conscious effort on my part.

I went outside and descended the stairs. Immediately upon reaching the bottom of the stairs, the force seized upon my whole body and I was propelled like a mechanical man down the street, halting at the corner. My body was spun around in a neat military right turn, and I walked to the end of the block. Here again I was stopped, spun in yet another right turn and propelled in this same manner right around the block. Returning to the entrance to my apartment, the force once again was removed as though it had been switched off.

Inside, I thought deeply about this entire process, not knowing what to conclude, since no communication of any kind had reached me other than the unintelligible scrawl on the book cover. All I could think was that this had been some kind of test. Today, of course, I look upon it as nothing more or less than *an attempt to gain complete control of me*, with disaster averted only by my absolutely inflexible resolve to remain master in my own house.

After a while I went to bed. Hardly in bed five minutes, I suddenly found my entire vocal mechanism functioning involuntarily, and through it, intelligent words and phrases were being conveyed to me. I did not hear voices, either at this time, or at any other subsequent time, but rather listened to my own speech mechanism, which was being used as the agency to communicate with me. The message as best I can recall, went something like this: "I am Ashtar. I greet you from Schare (Shuh-ree) your interest in our cause is well-known, and we would like you to help us. Would you be willing to do so?"

The message then conveyed some intelligence to me concerning a theory of mine on the UFO which proved my theory both incorrect and potentially dangerous to me. I therefore felt that this initial communication, containing such a warning with intimate knowledge of the subject concerned, was indeed genuine.

However, I was still greatly disturbed by the entire happening, even though I had sought to make contact with more than a little devotion. I had the inner feeling that all might not be what it seemed. Because of this, I sought out a kind gentleman greatly learned in occult matters, and related my experience to him.

From this man I obtained certain secret information concerning the protection of oneself during telepathic contact, since *all who seek to contact one by telepathy may not be spacemen.* This, my first contact with the world outside the purely physical, was a jolt to me. Reared in Christian Science, I was not prepared for any of these things, and least of all for the concept of aggressive forces in the unseen worlds.

These few vital minutes with this remarkable man were probably the most important of my life. Without the advice I received, and the knowledge to which I was made party in my need, I would no doubt have been in serious danger. At this point, as I see it now, although I could not see it then, I was about half ready.

CHAPTER TWO

"*FRIENDS*"

In the weeks and months that followed, I was able to appreciate that all beings, or intelligences, who circle our globe, or dwell near it, in high performance craft are not tall, handsome spacemen, free of dental caries and possessed of long golden hair. Nor are *all* these intelligences governed by a desire to uplift humanity. There are benign beings, and there are others, whose morals, ethics, purposes and allegiances are the antithesis of all that a Christian man holds dear. There are still others, not human, who are probably not even conscious of our presence.

The information that follows, and the comment on it by myself, is presented with the sole purpose of attempting to clarify the murkiest aspect of the UFO mystery for all who earnestly seek the answers. With the earnest seeking, it is almost mandatory for the truly sincere person to be intellectually bold enough to realize that the superphysical is very real.

No suggestion is made that all the information presented here is to be accepted as fact, for it is not so intended. Because of its origin, it should be subjected to the full glare of the reader's critical faculties. Every single statement must be turned over to the judgment, weighed, evaluated, and either cast out or accepted in accordance with what discrimination dictates.

The blanket acceptance of all alleged communications from "spacemen" will lead the student of ufology into a worse mire than the saucer writers have already made with their varying interpretations.

Let the intrinsic merit of the individual communication be its sole claim to belong in your personal world.

The messages are authentic communications, through a form of telepathy, by myself with an unseen intelligence or intelligences. In every case, I was fully conscious, and at no time was I under any form of control, seen or unseen. I had the power to instantly terminate any contact at any time, and the power to interrogate the invisible communicator in the course of the communications.

To forestall any objections parapsychologists may have as to the feasibility of such communications by telepathy, I state that these communications are carried on with the aid of an instrumentation system possessed and used by these invisible beings.

As students of hypnosis know, and as any good hypnotist can demonstrate, it is possible to gain access to the subconscious mind by placing the conscious mind in abeyance. By means of suggestion under these conditions the subconscious mind can be made to direct the action of limbs and perform tasks which in a fully conscious condition the subject would be completely incapable of accomplishing.

For example, Emile Franchel, well-known Los Angeles hypnotist, demonstrated on television one evening that a girl subject could write upside down and backwards with one hand and upside down and forwards with the other, *simultaneously*.

The subject had no ability to perform this feat in the conscious state.

In the use of this instrumentation system by these invisible beings, a similar state of affairs exists, except that in my case the conscious mind was not in abeyance, but functioning fully. The subconscious mind was apparently directed to operate the vocal apparatus during the time the beings wished to communicate. As to its feasibility, any person can establish for himself that a sympathetic movement of the larynx is inseparable from the function of hearing. No person should therefore conclude that any special ability is required. To the best of my knowledge, I have no telepathic power whatever with other human beings, and it is my present view that this *system* of communication used by these intelligences is simply tuned to the personal frequency of the human being and the intelligence passed in this way. This, however, is not the only way by which intelligence can be conveyed by invisibles. It is the method employed in my case, and one must either take this explanation or leave it as the discrimination dictates.

It would be extremely dangerous nevertheless, to say that it cannot be done. A short memory is the curse of a modern human, and we ought to bear in mind that only ninety years ago, a man was gaoled in New York because he said he could make what he called "a telephone".

These messages have assumed their particular character because of my insistence upon valid and useful information of high grade. Conditions of life on other planets, which obsess some people, hold no interest for me. Whether or not Mrs. Venus has a mixmaster or a washing machine is a matter of supreme indifference to me, but I am not indifferent to the presence, nature and meaning of the UFO. I have not sought prophecy at any time, although this work will presently contain one rather astounding example of this, but instead have sought explanations for various observed phenomena which have left our conventional science bankrupt.

The messages were compiled from notes made during the receptions. That is, during reception of the information, all pertinent material, names, and other portions likely to suffer from memory lapses were written down. Immediately upon terminating each contact the messages were compiled into written form as close to verbatim as possible. Later communications were taped, but none of these are now available and any tapes offered for sale using my name in this connection are bogus.

Through the years, there has been controversy as to the nature of the beings manning the craft we term the UFO. I prefer the term UFO because I do not believe they are all craft, and the photographs I have obtained bear this out. Theories, however, have ranged all the way from the obvious and elementary one that they are advanced beings from other planets, identical in shape and form to ourselves, to the almost shocking theory of Gerald

Heard, that only insects could stand such tremendous gravitational pressures and must, perforce, be the UFO crewmen.

Claimed encounters with physical beings in various places, where the entities have allegedly dismounted from saucers, have tended to support the simple theory of advanced humans from other planets. This is a theory that I hope this book will help overcome. Nevertheless, on the principle that anyone, including myself, could be wrong on any aspect of the UFO, I resolved that this particular aspect of the UFO beings should be resolved immediately:

Question: Are you etheric beings? Or are you possessed of a fleshly physical body such as mine?

Ashtar: "I am etheric. I do not have a fleshly body like yours, bounded by flesh. But it is possible for me to make my being visible to your optics by certain changes in its vibratory rate."

Question: "This would mean then, that you are normally invisible to us?"

Ashtar: "Yes."

If we take Ashtar at his word, we have explanations for some of the alleged encounters with space beings, and also for the fact that both ships and controlling intelligences are known to be invisible on occasions. These beings can convert to a vibratory rate where they are visible to us, but normally belong in a higher vibration.

It should also be carefully remembered that Ashtar is refering only to himself and those beings like him. We do not find any reference in his communication to any other faction which might be present among the UFO.

Question: "As you are an etheric being, are other etheric beings visible to you?"

Ashtar: "Yes, although not exactly in terms of optical vision as you know it."

Here we can readily see the difficulties presented to these beings in explaining themselves to a typical ignoramus in these matters such as myself. They can use only terms that are comprehended by us. It was apparent at this point that my own limited understanding was limiting the information that this intelligence could convey. A half-inch hose will only permit so much water to pass, and so it is with channels of information. Also, it is important to recognize that in communication of this kind, where the receptor remains in full and complete control of his faculties that the communicating intelligence can only use the words and expressions already present in the "mental telephone exchange" of the receptor.

It was about eighteen months after this communication had been received that I gained a rather startling proof of its validity. Realizing early in the game that an understanding of at least the rudiments of occultism and other aspects of philosophy was vital. I had pursued various phases of these subjects in the role of investigator.

I paid a call on a distinguished European lady in Los Angeles with an international reputation as a seer and psychic. Discussing various aspects of her faculty with me and answering my questions, she suddenly sat bolt upright in her chair. "Have you ever had any contact with any non-physical beings?" she asked, with a smile. I replied that I believed that I had, but could not be completely sure.

She then said "There is a magnificent looking being standing beside you, and he communicates to me that his name is Ashtar. Do you know him?"

I replied that the personality was familiar, and she gave me a full description of him as she saw him clairvoyantly. Seven feet or so tall, extremely stern, helmeted and giving the impression of being a sort of military man.

This was extremely interesting insofar as Ashtar had described himself as the "Commandant, Vela Quadra Sector, Realms of Schare, All Projections, All Waves," and therefore would probably be a military type of being.

Since I had not told her anything of my own activities in the UFO field, and she had never previously met me, I considered this incident to be a somewhat remarkable indication that Ashtar, in describing himself, had in fact told the truth, insofar as we could establish that truth. I saw nothing myself, being possessed of only ordinary sight.

There are many people interested in the UFO to whom the idea of a different type of being than ourselves is unacceptable. They cling valiantly and loyally to the idea that UFO beings are advanced humans from other planets and nothing else. These people want real, visible, physical UFO people, who will walk, talk move and *behave* as do our fellow men here on earth. This desire, born in many cases of lifelong religious beliefs, is likely to remain unfulfilled if what my research indicates is true.

The point that now deserves consideration is whether or not there is any evidence of a scientific nature that a higher form of mankind, or a different form of mankind perhaps invisible to physical eyesight, does exist? I submit that there is such evidence. There is evidence that every one of us possesses, in addition to his regular physical body, doubles of this body, formed of finer matter, which are normally invisible to the eye, and which interpenetrate the physical body.

For those who like to associate authority for statements with scientific qualifications, there are few men now living with more authority in these matters than Dr. Alexander Cannon. This great English neurosurgeon, psychiatrist, and philosopher, who possesses five Doctorates, has written a series of outstanding books dealing with the true nature of man.

On the question of this double of our physical body, which is termed in most circles the "astral body," Dr. Cannon has this to say in his book*"The Power Within:" "It is a duplicate of the physical body but composed of more tenuous matter which is invisible to the physical eye. Its organs correspond to those of the physical body, when active, and when passive it looks like a large egg of yellowish hue, hence the Biblical term (golden bowl)."

Dr. Cannon openly describes himself as "a scientist, not a spiritualist", and in accordance with this,

---

*"The Power Within" Sir Alexander Cannon, page 23-4. E.P. Dutton

goes on to outline a scientific experiment that can be performed to prove the existence of this astral body. "The astral body can be demonstrated to anyone by laying him in a bath of static electricity and holding an ordinary Osglim neon light near him, when the gas in the lamp which is held in the hand lights up, and one can definitely outline the astral body to even two feet away from the physical body; the static electricity does not light the lamp unless placed in the vicinity of the body of the patient, or subject, or an object." Doctor Cannon also postulates the existence of a still *higher* vehicle, the etheric body, and all this information is available to those who study his books. Not only does Doctor Cannon have a sound scientific background, including a plethora of orthodox qualifications, but he outlines these various truths about man with experimental backing. After reading his books, the Holy Bible takes on a completely new and different, and more intellectual meaning.

It follows therefore, that to believe in the existence of a real, but superphysical world does not place one in the company of nincompoops, but rather in the company of some scientists of exceptional skill.

In possession of just the beginnings of these general theories, I addressed further questions to Ashtar on this subject.

Question: "From your statement that you are etheric, am I to presume that you have evolved beyond the stage of a physical and astral body?"

Ashtar: "Correct. I do not possess a physical casing of the dense type such as yours. I am definitely etheric, *as are all people on other planets in this solar system.* However, this does not mean that we are invisible to each other as we are to you under normal circumstances. We see each other and live much as you do, but we do not have this dense physical casing which you possess. The advantages, benefits and comforts of this living are enormous, and the irritations of the fleshly envelope are most uncomfortable. Unless *we choose* to convert the vibrational frequency of our bodies to one which is visible to your optics, we remain invisible to your people. Highly evolved people, with a good "psychic eye" as you call it, can sometimes see us in vaporous form, although we may be invisible to other earthlings in the same location. When your clairvoyants travel to our civilizations on other planets, they see and are able to interpret our lives because they are not using their physical eyes but their astral or psychic sight, to which we are visible just as though we were physical."

Here then, is a voluntary offering by this invisible being, of the true nature of his particular faction among the many visitants to our planet. They live in a higher vibration than earth humans, in bodies designed to withstand this vibration, and which are quite probably similar to the fine, invisible additional vehicles Doctor Cannon shows that we possess. These bodies are invisible to the physical eyesight, but are nonetheless real.

If indeed we do possess these higher vehicles, and it can be shown that we have at least one such invisible body and possibly many of them, what is so illogical about supposing that one day we will function in these vehicles? Might not death be little more than withdrawal from the physical to dwell in the astral? These are weighty matters worthy of much consideration by those who seek the truth about interplanetary spacecraft, alleged spacemen and other flying objects which are neither of these.

In any case, these concepts begin to make the *nature of the reality* of these beings a lively one. They also begin to make many elements in the UFO drama assume a more proper perspective.

More questions along this line seemed to be in order.

Question: "When you become visible to our eyes does the person who sees you know that you are a "converted etheric?"

Ashtar: "Not as a rule. The conversion can be made so completely that a physical person encountering us thinks that we, too, are physical."

This would seem to assist in explaining many of the contacts with purported space beings in which the person experiencing the contact has sworn that these were physical beings. There is no question that such converted beings have appeared in some places to some people, and in some instances before witnesses.

The vanishing beings who appeared in a café to Mr. Truman Bethurum vanished instantaneously, a feat not performable by ordinary physical humans.

All these people experiencing actual contacts with these entities have risked having their sanity impugned by reporting the incidents. If they were to add to their reports the additional information that "these beings were not physical beings" then they would have been even more brutally crucified by the gentlemen of the press than they have been. It is probably not the wish of any *ethical* visitors to this planet, that persons be caused to suffer through these contacts, and hence, a "little at a time" has probably been their policy.

Convincing people has been a hard road, even when the physical line has been solidly adhered to, and there is no doubt that up till now a venture into superphysical science would have been casting seeds on stony ground. Men in general prefer to be confounded by a phenomenon than to have an explanation that they cannot understand, for what they do not understand, they hate and fear.

The preceding explanation naturally raised the need for a further one, relative to rides in saucers claimed to have been taken by some persons.

Question: "What of those who claim to have been up in your craft?"

Ashtar: "In our contacts with earthlings we have to be careful not to go beyond their understanding. In these instances, the ships and all entities within them are converted to a vibrational level at which they had the substance of physical things as known to you. Whether the experience was physical or astral is not known to some people who had the experience."

An astral experience is generally known in occult and metaphysical circles to be an experience which has taken place when the physical body has been asleep, and the "double" described by Professor Cannon leaves the physical and goes about its particular business. Rarely in the untrained person does a memory return with it, but in the case of these saucer rides, those of them that were astral experiences resulted in the person concerned retaining a very clear memory of the event. One which the person concerned could not separate or differentiate from the physical.

When one considers the knowledge of the mind possessed by these beings, as for example in the operation of the instrumentation system used to contact me, there is no doubt they could arrange for a memory to be retained if it was their wish.

It is also quite possible that many persons, perhaps *scores of thousands,* have been aboard these craft without being permitted to retain a memory of it.

Some instances indicate very definitely to me that actual physical rides have been taken aboard physical craft by some persons, but here again, we must beware of thinking that Ashtar and his particular group or faction represent the only spacemen or beings visiting this planet. The universe is a rather large place, and probably quite well sprinkled with assorted beings.

In this last communication much is explained by the simple phrase "we have to be careful not to go beyond their understanding." In the case of ninety-nine percent of the human race, such happenings as those just described are beyond the outer limits of the imagination. Generally speaking, we prefer to cling to what we think we know, and from this "secure" point to try and fit everything into the boundaries of established theories.

Needless to say, as a man of ordinary mentality, I was wallowing over my head at this point, and made the following observation to Ashtar:

Question: "The astral and etheric concepts are difficult to grasp at first."

Ashtar: "Persist and have faith. There is much to learn."

The last comment was a masterful understatement. Already a sense of my own puniness hung over me like a cloud. How would I ever grasp all these things? But, I resolved to try and learn more, since nothing had thus far been witheld from me that I was capable of understanding. It was back to ABC days once again, as I wiped the slate of my mind clean of orthodox dogma. I found that this aided the infusions of new knowledge that were offered.

My presumption now was that in this invisible world where they dwell, nonetheless real for its invisibility *to us*, they would have substances comparable to our own iron, steel, cement, and so forth, which would permit them to conduct the life which they had stated was similar to ours.

Question: I wish to ask a question or two concerning etheric substance.

Ashtar: "We will be glad to answer whatever we can for you. We wish to arm you with as much knowledge as possible, and are limited in this only by your power to assimilate it. I do not wish to talk over your head, but we will supply you with knowledge to the limit of your understanding."

Question: I am puzzled by the concept of etheric matter. For example, there is one case on record * where one of our jet aircraft flew right through a space ship without hitting anything solid whatever. Are your ships made of a vaporous substance, or are they a different form of earthly matter?

Ashtar: "We have all the elements you know on earth, and many more. The etheric form of these metals differs in its atomic and molecular structure from earthmade metals. For example, the distance between the nucleus and the orbiting electrons of the etheric iron nucleus is much greater than in iron as you know it on earth. This permits the atoms of earthly steel to pass right through the atoms of etheric steel in such a way nothing happens to either form of steel. The etheric form of steel enjoys a higher vibratory rate than earthly steel and therefore is not apparent to earthly vision or if you prefer physical eyesight. Under certain circumstances it becomes visible, as in the presence of certain atmospheric gases of Shan (Earth) or at will in accordance with the desire of the controlling intelligence. No matter how great the mass of the etheric substance, even a space ship measuring many miles across in your measure, physical matter cannot damage or injure it or its contents."

From this it would seem that the practice of shooting at UFO instituted by the various air forces is

---

*Capt. William Maitland, U.S.A.F., Chicago Ill. September 2,1952

wasted effort. At least this is so relative to these etheric beings, and here again we must remember that there are quite possibly others, a point we shall cover directly. Aside from the fact that the UFO entities seem able to read the minds of our pilots with terrifying exactness, as revealed in the published accounts of such encounters, even if the jets were in a shooting position, the shells or rockets would have no effect upon etheric craft or beings. They are, in fact, a different order of matter than our own.

This information has merit insofar as it is an explanation of the peculiar case where this jet fighter did actually fly through the saucer, and was seen to do so by ground radar.* What official explanation is there for this, aside from the usual childish gobbledygook released by the air forces?

Furthermore, this explanation of the nature of the etheric substance involved in some of these craft, together with the idea that it becomes visible under certain conditions, accounts for many sightings. I refer especially to those sightings involving light manifestations and vaporous objects, rather than clear cut, solid, physical things.

Question: When you speak of making etheric matter visible at will, is this the way that George Adamski was permitted to take his now famous photographs?

Ashtar: "Yes. Ether ships as they have been called on your surface, have been made visible to and for certain individuals, selected, upon your surface of

---
*Captain William Maitland, USAF., Chicago, Ill.

whom Adamski is one. *Normally,* the ships are part of the *invisible* world."

This latter statement of course, is at variance with the theories of Mr. Adamski, who claims that the whole manifestation is one of physical beings ushering in a great new scientific age. Be that as it may, there is no doubt that Adamski's photographs remain among the most startling evidence yet obtained of the presence of the UFO. I have no desire to enter into any controversy with Mr. Adamski or with anyone else regarding his role in the UFO puzzle. I wish him, and every other investigator, nothing but the best. Harking back to the information given regarding a special, extended or different type of vision possessed by some people, it seemed to me that development of this "astral" vision should enable one to "see" the craft.

Question: If one were to develop astral vision, or the psychic eye, would he be able to see the ships?

Ashtar: "No. Not unless the vibratory rate of the ship were converted to the vibratory range of astral vision. Remember, the etheric vibratory rate is higher than the astral. Very few physical humans have some perception of the etheric, but they are not normal people as you know them and for the most part dwell in very secluded places. As a general rule, perception of the etheric through vision cannot be accomplished except through the will of the etheric, converting etheric substance to a vibratory level where it is physically visible."

I wondered if I would ever be able to chew some of this material I was biting off in such large chunks.

It was at once fantastic and feasible. Its application to UFO sightings caused them to begin making sense The explanation of etheric matter and its properties and behaviour, *eliminated puzzlement,* and that is after all, what we are seeking to do by investigating the phenomenon.

The next manifestation on which I sought information was the "fireball." Most puzzling of all the UFO phenomena, the fireballs have been seen in various parts of the world during the past eleven or twelve years. They have varied from a harmless burst of fire that leaves no trace, to searing missiles that have hurtled through hoardings and buzzed around bedrooms leaving fire, destruction and consternation in their wakes.

Question: Are these fireballs dropped from your craft?

Ashtar: "Yes."

Question: What is their purpose?

Ashtar: "The fireballs are nullifiers for certain radioactive effects which your scientists do not even know they are releasing. These nullifiers prevent the poisoning of your people, and are released upon the orders of our Heavenly Father. We cannot intervene actively in the cessation of such experiments but we must do all possible to prevent the wanton destruction of human life resulting from these experiments."

This seems to be a fairly logical explanation, even though no attempt was made to explain the nature of the unnamed radioactive effects. Probably such information would be very difficult to convey to a non-technical person such as myself, and quite possibly no words existed in my "mental telephone exchange" to enable them to give me an explanation.

A little less logical and more likely to be treated skeptically is the idea that the Heavenly Father would personally intervene in these matters. This will remain a matter of personal opinion for some time but I do not entirely subscribe to the idea that the Great God of the Universe, to whom Ashtar apparently refers here, would set aside more weighty matters to personally order such activity.

In this, as in all else, time will tell.

Question: It has been noted that the air contains a large percentage of copper after these fireballs have been seen.

Ashtar: "This is true. It is part of the fireball's functions. For the most part, these fireballs are seen in the vicinity of atomic installations and laboratories. But *not all fireballs are ours,* and all do not have the *same purpose.*"

This final sentence set me to thinking very hard along the lines expounded by Mr. Harold T. Wilkins the distinguished English UFO writer. Mr. Wilkins' first UFO book, bearing a title thrust upon him by his publisher, without his consent, disturbed many saucer fans. It was sold as "Flying Saucers On The Attack." This caused upset and consternation among saucer fans who could not bear to think of the saucer entities as not all being beneficent, handsome beings intent on saving mankind.

Nevertheless, several mentions made in his book by Mr. Wilkins of weird craft starting fires seemed to mesh neatly with this particular communication from Ashtar.

The implication of the message is clear. More than one faction must be at work, for "not all fireballs are ours." Soon I was to learn the truth of this, and in learning this truth, to gain information which unravelled many of the unsolved mysteries of the UFO.

I pondered upon this introduction of additional factions among the UFO. I pondered upon the duality of life as we know it here on our earth. We have male and female, positive and negative, and up and down. We also have good and bad, although these particular qualities are measured by the individual and not in absolute terms. We also have benign and malevolent, two opposites of similar nature to good and evil.

I could not see anything preposterous in supposing that this observable, measurable duality of life which we know in our own world, would apply in varying degrees to other worlds, whether we could see these worlds or not.

On the contrary, it appeared to me to be absolutely logical that this duality would be preserved, and that if we could but perceive it, our understanding of these weird visitors to our earth would be improved.

CHAPTER THREE

FOES?

And he laid hold on the dragon, that old serpent, which is the Devil, and Satan, and bound him a thousand years.
And cast him into the bottomless pit and shut him up, and set a seal upon him, that he should deceive the nations no more, till the thousand years should be fulfilled and after that, he must be loosed a little season. —Revelation 20:2-3

At about the same time I received the information concerning the fireballs, especially that portion which stated that all did not come from the same craft, I came into contact with the now famous "Shaver Mysteries." In these writings, Richard L. Shaver claimed that certain entities, by an unique process spoke to him through his welding machine. He further claimed that these entities identified themselves as the "dero," an underground race living today within our planet.

There is a ready tendency to laugh at the idea of entities using a welding machine to contact a human being. I find nothing humorous in it, rather do I find it logical, for my researches indicate today that these various entities require the energies of *this plane* of life in order to manifest *here*.

Aside from this, everything deserves investigation that seems to tie into the UFO mystery. Because of this viewpoint, I resolved to ask Ashtar what truth there might be, in his opinion, regarding this claim of an underground race.

Question: "A question that has greatly disturbed me, concerns the presence of an underground race on this planet. I would appreciate any information you can give me."

I was hardly prepared for the import and impact of the answer I received. In one stroke it minted a key to the UFO mystery, and dispelled much confusion.

Ashtar: "At the *core* of your planet, there dwells a greatly degenerated race, an astral race, which is degenerate not so much in science, but in every moral respect as you know and understand it. They are capable of space flight within the astral regions around the earth, but are earthbound. They are the forces of Eranus, whom you call Satan. They emerge at the South Pole. On your surface, they have allies who are without morals and without mercy. I give you this information that you may be aware of their existence. I enjoin you to forever close any researches into this astral activity, in the interests of your own safety.

Be on your guard always, be careful and vigilant."

To suppose that I was anything less than shocked by this information would be to attribute to me more raw courage than most lions possess. I was aghast and appalled, and not a little confused by the information, and at first almost dismissed it as cosmic science fiction.

However, a careful review of the communication makes the puzzle more like a picture. First, he speaks of the *core* of the planet. It is important that we remember that he is speaking of the core, and not of any *intervening* levels of life that may or may not exist. We shall return to this shortly.

Secondly, an astral race is mentioned. This would be a race of finer, differently vibrating beings than ourselves, invisible to us except at certain times and under certain conditions. Degenerate not in science, no, for do they not whisk through our atmosphere in craft of high performance? This was beginning to make sense! Degenerate in every moral respect as we know it. When we review a little later some of the very unsavory manifestations connected with the UFO which have been suppressed by the censors and glossed over by the saucer fans themselves, we shall see that this statement about their morals has exceptional force and merit. The forces of Eranus? Who's he? The being we call Satan! Is there in fact such a being or deity? Could it be that the Bible is not after all the pure trash some scientists regard it as being? Could it be that some of the UFO constitute the air force of this dark deity?

I was soon to have the name, origin and nature of Eranus brought to my attention under exceptional circumstances, and I already felt that I was on the right track. I was beginning to get those first vital glimmerings of duality, the bi-polar manifestations that make up the whole!

They are said to be capable of space flight, these beings, but only to a limited extent, according to this communication. They are earthbound, and apparently unable to penetrate beyond this earth's "envelope." This envelope does not seem to coincide with the atmosphere, but rather to extend a considerable distance into space.

If these forces "emerge" at the South Pole, this could partially explain the feverish military activity at the South Pole. Aside from America, Australia, New Zealand, Britain, Norway and several South American nations have great interest in expeditions operating at the South Polar regions. Even the Russians, whose interest in the Antarctic has been virtually non-existent since a Tsarist naval expedition attempted a mapping of the Barrier, are down there in force.

The Soviets, in fact, are establishing a line of bases on their way to the "Pole of Inaccessibility" the area between the South Pole and South America. This part of the Antarctic embraces an unexplored and unseen land area half the size of Europe. It is reported now that the Soviets have discovered in this locale, an ice free area, with brooks, moss and lichens some 4,000 feet *above sea level*.

It is true that some of this Antarctic activity can be explained away by the International Geophysical Year, but the endeavors of all these antarctic groups would seem to be aimed at something *much more permanent*.

In the July 1957 issue of "Fate" magazine, commentator Frank Edwards refers to the experience of

Chilean Naval Commander Orrego in antarctic waters. His squadron of ships was circled by UFO's which seemed to vanish inland over the ice after they had thoroughly surveyed the Chilean naval vessels.

The connection here is very important with the many reports coming from Southern Hemisphere countries indicating that extensive course plotting has been carried on by the military authorities. The paths appear to lead to and emanate from the south, leading to the conclusion that the origin of at least some of the UFO is a south polar one.

New Zealand sightings and records, with which I have had considerable contact, contain large numbers of sightings of cigar shaped craft, many of them involving *blue emanations,* heading toward or coming from the South.

And then there is the opinion of Commander Justino Strauss of the Brazilian Navy, who recently wrote a description of the interior of the earth for a Brazilian periodical. He described it as a world of great beauty, but illuminated with a sort of *pale blue luminescence.*

We do not have any authority for this description, unless Commander Strauss has made the journey, but the fact that he is a Navy Commander should indicate that he is both rational and responsible.

In judging these viewpoints critically, it should be borne in mind that it is unlikely that we are going to suddenly stumble upon a vast aperture in the earth near the South Pole, through which these machines emerge. If we accept the concept that they are of a different order of matter to our own, we may expect that their matter will probably interpenetrate ours at the Pole if not elsewhere. If indeed, they are of "astral" character, then the indications are that their matter will interpenetrate ours, in the same way as our astral bodies interpenetrate our physical bodies according to Dr. Alexander Cannon.

If, on the other hand, three dimensional craft whose possible point of origin is our moon, are frequenting Antarctic surfaces, we might expect to find such an entrance. The term "emerge" may also be synonymous with "materialize."

It is my *belief* that the South Pole, or some point near it, is used as an entrance for this subterranean lair because the magnetic propulsion systems of the craft require that they proceed along the pole to reach whatever type of neutral exists at the core of the planet, where they are purported to dwell by Ashtar.

As for Ashtar's final injunction regarding closing researches into this activity, I needed no second suggestion. In my view, if these beings were as stated, then clearly I was not in any way equipped to deal with them. If, in fact, I was to stir them up by sticking my nose into their business they might have a power of retaliation that would cause me to rue the day I was born.

The information seemed to have been given as a key to at least part of the UFO mystery, to dispel confusion, and above all, to convey to well-meaning but naive and spiritually helpless people the fact that all of the UFO are not as rosy as they seem!

Here was the first element of duality, the first trace of logic, the first evidence that perhaps these other worlds contain good and bad just like our own.

One manifestation seemed to be of a very high ethical origin, describing themselves to my limited understanding as etheric. These beings were assertedly of some kind of interplanetary force, probably sent to the environs of Earth as guardians.

The other manifestation was probably what the guardians were needed for. Expeditionary forces are not normally sent out without a job to do, and probably this second manifestation has become virulent and needs watching. This latter manifestation purported to be of a lower order of matter than the first, described to my limited understanding as astral, possessed of ethics and morals of a satanic nature. This is described as being *of this earth and confined to it.*

Unfortunately, the "dero" had not yet been mentioned and at the next opportunity I pressed for a clarification on this point. I was not satisfied that the information already given entirely fitted the "dero."

Question: These astrals from the earth's core are, I take it, the "dero" of legend?

Ashtar: "No. The Dero are no longer confined in the caves of legend and story, but are reincarnated upon your surface. Many of your eminent scientists, driving forward with the perfection of ever more prodigious blast forces, are reincarnated deros. Because of their prior lives as cannibals and degenerate beings, they are prone to exert themselves for destructive ends, since they fall easily under the astral influence."

I do not wish to enter a dispute with Mr. Shaver, nor to question the authenticity of his statements I have every reason to believe that he is a truthful man. The above information from an unseen source however, would seem to check with what we know to be true about certain scientific minds.

Some of our scientists have gone through moral torture of a type not comprehended by the average person, when they have had enough moral fibre to review the monstrosity of their handiwork. Many of them have turned away from destructive work and refused all inducements to return to it. These men are worthy of honor.

There are others though, who seem to function best when perfecting some newer and more prodigious weapon for searing and sundering their fellow man. It would seem that such men are under the influence described. I have met some of them and have noted their complete lack of any spiritual life. Many of them dismiss religion as "priestly potash." I wouldn't dare to judge these men for all beings have a role to fulfill and a right to live their lives in their own way, I merely note that they are there and that they fit the description offered by my unseen communicator.

At this point we might mention the stories related by Gray Barker in his "They Knew Too Much About Flying Saucers," concerning black suited men. These worthies apparently paid unscheduled visits to saucer investigators with the idea of shutting them up when they got too near the truth.

In my view, some of these investigators may have stumbled on this center of the earth concept, probed into it, stirred up these forces and consequently been handled harshly by their earthly representatives. These black suited gents would probably be the "allies on your surface" who are without morals or mercy.

Only one of these black suited boys was ever identified properly, and it was one who was seen sitting in a car down in Australia where some of these unsavory visits took place. This particular man was a known criminal, wanted by the police. There is, of course, no more fitting servant of Satan than the earthly criminal, or a member of our earthly underworld. These beings occupy themselves in left-handed endeavor and naturally are servants of the forces of the left, or darkness.

In the various reports of these visits, which could hardly have been by any form of Government men, the nature of the boys in black is clearly evident. They endeavoured in every case, with threats and exhibitions of power to fill those they visited with fear.

In some California cases, and also in Australia and New Zealand, attacks on animals using strange powers were made by these men in black. They must be typified by such dastardly acts.

Men who call and frighten and threaten men with families, men who are known criminals, and men who do not carry authority of properly constituted governments ought to wear black suits, for these black seamless suits are themselves their badge of allegiance.

Their chief weapon has been *fear*, the favorite flail from the Satanic armory.

The greatest of all men ever to tread this globe, Jesus the Christ, left us an all-time yardstick for such personalities. "By their fruits ye shall know them." The fruits of these men in black clearly show how they are to be known, and let us so know them.

The right to life is possessed even by such entities as these, and danger lies for those who want to go around destroying "evil" things, or in poking into the business of these dark personalities. After all, the average person who is interested in UFO's, perhaps investigating them, wouldn't dream of trying to investigate the underworld in New York City for example. Yet, there is a similar side to the UFO which is best left to those who not only know *how to deal* with these forces, but are *authorized* to do so.

This is where these so-called etheric beings probably fulfill their role. They are the counter manifestation, the one whose plan of action has been laid where plans do not misfire. There is, therefore, no reason for people to become fearful or disturbed about these forces provided they recognize the right of them all to exist.

All this astral and etheric idea seemed to be working out fine, but there still seemed to be far too many instances of physical craft that remained unexplained. It was time to ask again.

Question: In our solar system, are there any other physical beings like us?

Ashtar: "No. All beings on other planets in your solar system are etherics. On your planet, as you now know, there are two kinds of beings, physical and astral. Outside the earth-moon system in your solar system, all are etheric."

The major portion of this statement is unprovable for the present, but the implication is clear that physicals also exist on the moon. By this, we mean fleshly type human or humanoid beings of similar order to ourselves. And the presence of physicals in other solar systems is not denied.

The presence of physical beings on the moon does appear to be fact. The actual signs of construction present on the moon, in material of a vibration visible to us, clearly indicates that someone capable of fabricating physical matter does live or is based on the moon.

Relative to these terms astral and etheric, which are loose terms, it should be pointed out to potential critics that these are very broad classifications used to permit me to grasp the differences between some of these various beings connected with the UFO.

Conversations with learned people who are occult science students have indicated to me that this is nothing more than a rudimentary, easily grasped explanation, especially employed for the benefit of my limited understanding and knowledge.

No one should consider these terms "astral" and "etheric" as being universal and all-embracing, but rather as a guide in mastering the truth about the UFO. As with most things, framework first, details later.

The idea of limited flight capability had already been mentioned by Ashtar, and I felt that it might be as well to get further information on just how hot these hostile space pilots were. It seemed that they were not as hot as they had led some people to believe.

Question: How far from the surface of the earth in our measure, do the astral regions extend?

Ashtar: "125,000 miles. Within them, the astral beings are confined. At certain times of the year, travel to the moon is possible to the astrals when the astral shells of the two bodies overlap. When these two shells separate however, any entities on the moon are cut off from the earth until the next time the astral shells overlap. *No physical or astral entity can go beyond the earth-moon system.*"

At this time I was not familiar with the term "Perigee" of the moon, and hence this could not be conveyed to me. But it is a safe presumption that the perigee of the moon is the time when this commerce takes place.

The probability is very great that to our present earthly observational equipment, any craft capable of travelling 125,000 miles into space, or to the moon, would be considered a "space" craft. If this communication is true, these "astral" craft can attain this altitude and no more, except at the times mentioned. The pattern now began to assume more logical outlines.

There are quite possibly at least two types of craft, one a genuine interplanetary craft, capable of travelling through space at will. The other, a pseudo space craft, with a definite performance and altitude limitation. We can tentatively assume that this latter type of craft is malevolent, and probably responsible for many acts of violence against our surface carried on through the ages. They are quite possibly of Satanic origin or allegiance.

The interplanetary craft, or constructs, termed etheric, or etherian, and apparently controlled by beings who are also of a high ethical nature, represent a counter manifestation. They are our positive pole, so to speak, in our search for duality in the phenomenon.

Despite everything, the idea of three dimensional craft still stuck firmly in my mind. Could there be a third faction or a fourth or a fifth? And harking back to Antarctica, rumors have made the rounds, of craft and cities in this area. This may not be as ridiculous as first assumed, especially in the light of the discovery of an ice-free area by the Russians. Another question was in order.

Question: Three dimensional craft have been seen in addition to the others described in these contacts. They are believed to be of a mechanical type. Are there such ships, and where do they come from?

Ashtar: "Yes, there are such ships. They come from the Continent that you call Antarctica. Before the Lemurian flood and rebalancing of the earth on its present axis, this was a great civilization. It was frozen quickly and its splendor swallowed under the ice. One city, Rainbow City, remains above the ice with a few dwellers. They have these craft, and there are also some in Tibet. These dwellers are entirely separate from the forces in the center of the earth, and plan hostility against no one. Mechanically propelled, these craft also outperform earthly aircraft."

There is no more dangerous ground on which an investigator can venture, than the shifting sands of Lemurian or Atlantean legends. This communication, at least in this respect, remains unprovable for the present. The expeditions in the area of the South Pole however, may one day send back the startling news that they have found these things. It is a fact, however, that the South Pole is a center of UFO activity, and this cannot be lightly discarded.

On the question of Tibet as the point of origin for some of these craft, this planet's population of technological lunatics can be expected to emit resounding guffaws.

The objection first registered is that Tibet has no industrial base from which such devices might be derived. This is true. Scientists who have been to Tibet however, find that many wonders are wrought in that remote and sombre land by powers that confound the most astute scientific minds. A heavy industry may not be at all necessary for unlocking some secrets of space flight, and the rational investigator is entitled to ask why Communist China has sent its brutish swarms to overrun such a barren country. Could it be that China's ambitious and aggressive rulers are seeking to force from Tibetan guardians of these special powers secrets that could lead China on broader conquests? It is possible, and perhaps more than possible.

The most interesting light yet shed on this question of Tibet as a base for spacecraft operations comes from Dr. T. Lobsang Rampa, author of the book "The Third Eye." This strange writer in a purported lifetime of training in Tibet, saw many things Western eyes are unfitted to behold as yet.

In an article in the May-June issue of the "Flying Saucer Review," Dr. T. Lobsang Rampa describes a personal visit by himself to a remote and lush valley high in the Tibetan wilds used as a base for many different types of spacecraft. He details his visit aboard such a craft, and a flight. It is possible he speaks the truth.

Physical craft visible to the eyes as solid objects are more readily acceptable to the investigator than invisible craft. However, what the investigator is prepared to accept and what the truth is are not always the same thing, so one must investigate as an open-minded skeptic.

The "astral regions" of which Ashtar spoke, seemed to me with my skimpy knowledge of such matters, to be inseparably bound up with spooks, discarnate humans, spirits and similar things. In this terrain, I was most definitely a stranger, with neither helm nor compass. Compounding this situation was the fact that my lifelong religious training in Christian

Science had always stressed that such things were nonexistent and unreal.

However, in spite of these taboos of childhood, this phenomenon of the UFO did seem to be leading into the realm of spirit and spiritual phenomena. More information was called for.

Question: What is the nature of the astral regions around the earth?

Ashtar: "The astral world is divided into two broad sections. First there are the bodiless entities from your surface, the so-called dead people, who must become incarnate again in order to pass completely to the etheric state. Some of these entities are waiting what will be their last incarnation. Others are those who have had their carnate existence terminated abruptly or accidentally, such as criminals and soldiers. All these entities have in common the intense desire to become fleshly once more, in order that they may qualify to be no longer earthbound, when their incarnations terminate. This is the Garden of Waiting. There are also the monstrosities and phantasmagoria which are degenerate thought forms. The other great section of the astral world is the astral regions of evil, which surround and interpenetrate the earth, inhabited by beings who are forever discarnate and forever earthbound by decree of the Great God of the Universe. These beings cannot enter the Garden of Waiting. It is against these forces that we of the etheric world are warring."

Persons highly trained in esoteric philosophy will doubtless find much in this communication that is debatable. However, to me at the time, it implanted the idea of astral beings who are unable to incarnate, and unable to pass on. Beings under a penalty of some kind. And if, as Ashtar claims, these beings had been placed in this state by the Great God, surely their crimes must have been of an exceptional nature, to so provoke the Great One.

When this communication first appeared in my earlier booklet, a storm was stirred up by various sects and individuals who found it at variance with their own particular concepts. I have already mentioned that these communications are broken down and simplified by my communicator because of my limited knowledge. They are neither exhaustive, as most occultists recognize, nor are they forced upon people. One must accept or reject what the critical judgment dictates. One gentleman who wrote me suggested that if I would change the text of the communication he would be prepared to accept it. He outlined the changes he felt were necessary. He just expected that I would change the Ashtar communication to suit his views.

The major objection registered to this communication lies of course, in the imposition of these penalties on certain of these astral beings. "Forever earthbound" "forever discarnate," for example, were felt by some people to be too severe penalties to have been laid on by a God of love.

Subsequent inquiry on this subject to Ashtar merely reaffirmed this portion of the communication however, with the only addition being that deliverance could come to these beings only if they sought pardon and mercy at the hands of the Almighty. Then, it was stated, His mercy would be manifested. Until that time, these forces remained essentially *untamed spirit beings*. Ashtar could not be drawn into explaining the reason for their predicament, or for their relegation to this invisible domain around the earth. It seemed as though the door was slammed shut. My answer was to come through another channel.

Not long after the receipt of this communication about the astral regions and their occupants, I was looking through the shelves of the New Age Publishing Company's bookstore in Los Angeles. I seemed to be drawn to a certain book, which I took down from amidst many hundreds of books on the shelves. As I took it out of its place, it fell open in my hand at a particular page and I felt impelled to read right at this point. The effect was electric! I had been guided in some way to a passage which explained the very thing that had puzzled me.

The book, "Atlantis To The Latter Days" by H. C. Randall-Stevens* turned out to be a history of the beginnings of the earth given to Mr. Randall-Stevens by clairaudient dictation. The invisible dictator of the material identifies himself as Osiris, son of El-Daoud. El-Daoud, with his female half Evam, is alleged to have been placed in charge of the evolution and ordering of the material Universe by God.

At this point, we are getting beyond the fringe of the occult, and into matters which are not viewed as the province, interest or business of present day materialist thought. However, Mr. Randall-Stevens' communications are entitled to be heard with an

*Aquarian Press, London -

open mind, for in earlier days this same invisible Osiris gave Mr. Randall-Stevens information regarding ancient Egypt that was not at the time a part of accepted theories. In particular, the invisible being controlling Mr. Randall-Steven's arm, made an exceptionally fine drawing of certain portions of the Sphinx which had never been excavated and were not known to exist at the time.

At the time of its receipt, the drawing and the information were both given the intellectual bum's rush by case-hardened archaeologists. However, later excavations in the appropriate sections of the Sphinx and Pyramids revealed things to be exactly as drawn by the invisible being. The archaeologists sat down to a hearty meal of crow.

Here then, is the point. Some of Mr. Randall-Stevens' previous communications have been vindicated by physical discoveries that are now scientific fact. Because of this let us lend a critical ear to what this once vindicated Voice tells us about earlier times on this planet. Let us see how it fits with what we know, or what we think we know, about the UFO.

Here are the relative passages from this remarkable book. It is Osiris who dictates the following to Mr. Randall-Stevens.

"The first Son-Daughter of the Father-Mother-God was even my beloved Father-Mother, El-Daoud-Evam. El-Daoud-Evam are set down in your Bible, Oh, my children, as Adam and Eve. Later you will see how the Bible came to be distorted and altered out of all recognition from the actual laws of the Divine Will of God It is necessary that I briefly outline to you the descent of certain of the Ray Dhumans from the Father-King who is even my Father El-Daoud. As I have already told you, El-Daoud-Evam was the first Son-Daughter outside the actual trinity of Godhead. My children, El-Daoud was then given certain instructions by the Godhead. He was told what were to be his duties, and those of his Spouse-Eternal, and he was also told that he would be placed in charge of the evolution and ordering of the material Universe of God — that part which was to comprise the grosser forms of matter which were the furthest removed spheres from the Godhead, and which were fourteen in all, when they were created. So, my beloved children, this was the great work of my Father El-Daoud, that he should be the Father-King and Viceroy of the Godhead unto the material spheres of which Earth is the grossest and furthest removed from the Godhead..."

The story continues, relating how additional sons were brought forth by the Godhead to aid in the evolution of the Plan. One of these was Eranus, whose name is already familar to us from Ashtar who described him as the being we call Satan. The other was Yevah, whom we could presume has come down to us as Jehovah. Thus, these two were Sons of God, and brothers of El-Daoud.

These two divine sons of the Cosmos, deities in their own right, committed a series of crimes in partnership, according to Osiris, dictating to Randall-Stevens. They had been appointed Viceroys of the Earth's evolution, but Eranus, the subtle one, draws Yevah into his plans for accelerating the evolutionary process. The materialization through their God power, of mortal sex bodies and assorted monstrosities is described. Of these two deities, Yevah repents of his crimes, but not Eranus, who continues his incredible transgressions against the Father-Mother God. Osiris relates these events thus:

"Alas, all reason was useless, Eranus, who shall henceforward be spoken of in this my book unto you as Satanaku (Satan) defied God the Father-Mother, and pitted his will against the Divine Will of Godhead. He was fully awake to the possibilities of his own Divine powers, and he intended to use them, and to use them in any shape or form against those that stood in his path."

The close correlation between these communications and those from the being named Ashtar is clearly seen. We are actually in the process here of finding an explanation of the forces of the Left, the forces of Darkness, who they are and where they are. The decision to accept or reject lies of course, with the individual.

Continuing his narrative by dictation, Osiris relates the story of yet a further series of shocking crimes which seemed to me to be aimed at the Creative Force or, a direct attempt at usurpation of the power of God himself. The results of this second series of transgressions are described on page 114 of Mr. Randall-Steven's book:

"Satanaku was now beyond all reason whatsoever, and so was confined to the workshop which henceforward was called The Astral Regions of Evil, where his powers of interference were very much curtailed. But, alas, his will to thwart the plan of Godhead, and bring shame and ruin on the name of El-Daoud and his Twin and Spouse Eternal was strengthened one hundred fold. My children, remember that God never requires the death of a sinner, and even as was Satanaku, the Prince of Evil, allowed time in which to repent, so are all of you who have in like manner done evil in the sight of God, the Father-Mother. Also, God said unto El-Daoud, 'I will not destroy evil, neither shall I utterly limit the power of Satanaku, for by means of that evil which he hath created by the misuse of Free Will, my Divine gift unto the Dhuman-Adamics and Yehavics alike, will I test those men of earth who have to evolve upwards that they may become my children by adoption. And I will cause those men of Earth to be strengthened by that very evil which has been created by the will of Eranus who is now Satanaku, the Prince of Evil-Doers, who together with his co-workers, shall be confined to the workshop of Earth, which place shall hence forward be called the 'Astral Regions of Evil'. And he shall remain there until that day shall dawn when he shall again come unto me seeking pardon at My hand's."

These events took place first in Lemuria, scene of Satanaku's earlier transgressions, and later in Atlantis, preceding his confinement to the "Astral Regions of Evil."

In these communications there is some support and endorsement for Ashtar's communications on the same subject. There is also a remarkable similarity in terminology in the case of the "Astral Regions of

Evil." If we are to believe the Randall-Stevens messages, this was in fact the term ordained for these regions by God himself, and spoken by Him as related above.

We see from these communications that so merciful is the Great God that *even Satan* has time to repent.

Meantime, what is so illogical about Satan being confined to an area where his power will be limited, although not utterly? His evil power, and his evil works are used by the Almighty to test men, and who can deny that we grow in spirit as we overcome evil? And if there were no evil, how could we measure good?

In addition, it would be completely out of keeping with the patent order of the Universe for such untamed and violent beings to roam through it at will. What could be more logical than that the Almighty would limit the power of, and confine a lesser god, who had sought by every evil device and perversion of his god power to usurp the power of the Almighty? In such matters, we might expect that the Lord God is a "jealous God." And we should remember, as additional support for this theory that the Commandment is plain: "Thou shalt have no other gods before me." Presumably this includes Satanaku, a lesser god.

Elsewhere in "Atlantis To The Latter Days," Osiris states that the very atmosphere around the earth constitutes part of the astral regions of evil. The Bible also refers to Satan as the "Prince of the

Powers of the air" and also states that he "shall be bound."

When airplanes go aloft into the atmosphere and carry humans within them, they are possibly intruding into the domain of these beings, if we regard these communications as being valid.

Intruders are not welcome at any level of society, and it is no bold supposition that they are not welcome in someone's "workshop". We shall presently see how this concept meshes neatly with the known evidence, evidence that clearly indicates to anyone unafraid of giving the UFO a spiritual and spirit aspect, that among the UFO there are both friendly and malevolent beings, as well as others who may be indifferent.

In the last Ashtar communication, he stated that they were warring against these evil forces. The nature of this conflict would be of interest, perhaps, in explaining other observed phenomena:

Question: What form does this war take? Is it a clashing of space ships in combat?

(Note the childish nature of this question of mine, and marvel at the patience of a highly developed intelligence to whom it was directed)

Ashtar: "It is not a matching of violences, as you suggest, but a *battle for the control of earthly minds.* Our purpose is to overcome the destructive influence, the physically violent influence which the dark ones seek to exert over mankind. Our purpose is to nullify the astral influence by restraining beings devoted to destruction and physical violence. The dark ones seek to relegate the whole world to the darkness wherein they dwell and have power, and thereby increase their influence further. Our task, as decreed by the Heavenly Father, is to nullify, overpower and banish the work of the dark ones by good influences upon humanity. This is the true nature of the battle, rather than spacecraft versus spacecraft."

In the first place, if we are to accept the statement of Ashtar that his forces are etheric, they are probably not entirely beyond damage at the hands of the astrals, even if they are of a different order of matter than the astrals and their craft. It would be unlikely that the dark forces would be able to strike at them at will, even as it is probably impossible for us to strike physically into the astral world, or "astral regions of evil" other than with the atomic bomb.

There can be little doubt in the minds of intelligent persons that certain human beings, and certain nations, permit themselves to become the tools of evil forces. Perhaps in due time we shall see how this see-saw between the forces of Right and Left, Light and Dark, extends into our contemporary world here on earth.

Along with much information that I regarded still with some skepticism was this question of the center of the earth, the purported abode of these satanic forces. Although I had been enjoined to leave well enough alone, I resolved to ask about the nature of their dwelling place.

Question: What is the nature of the core of our planet? Is it solid? Or hollow?

Ashtar: "The center of the earth consists of matter of a density comparable to air, although it is not air. You would term it hollow in your expression. It is here that the forces of Satan dwell. Near the South Pole they emerge in their craft and circle the planet. Clumsy and primitive by our technology, their craft are still greatly advanced over yours, and they are easily able to outperform and to outmanoeuver mechanical aircraft of physical manufacture. They are also considerably faster, being capable of speeds in excess of three thousand miles per hour."

It is interesting to note that for ages the legend has been that Satan is in hell and that hell is down there, underneath. Some Tibetan teaching speaks of a "cold hell" however, and this might apply equally to the South Pole, and to our own atmosperic envelope, relative to the temperatures we know and regard as hot and cold.

And has the reader ever watched a horror movie and felt the chills run down his spine?

Perhaps matter of comparable density to air is what we have at the earth's center, and for those who fancy their personal hell to be a hot one, sulphur dioxide created by burning sulphur has a density comparable to air. Also, it is not air, as anyone who has inhaled it will testify.

All theories and indications are, of course, that the center of the earth is as "hot as hell," but the time-honored one that it is a molten mass is pure theory. No one has been there to find out. The deepest oil bores and mines are but infinitesimal fractions of the total thickness of the earth. Since we don't *know* what is down there, one theory is as good as another, and Ashtar's description may prove in due time to be correct.

In connection with temperatures, vulcanologists and geologists might take a word of warning from investigators of the upper air, who labored for decades under the misapprehension that the higher one got the colder it would be. Temperatures have turned out to be somewhat stratified, and not progressive, in those levels of the atmosphere they have thus far investigated. Again, many German U-boat sailors are alive today because in the depths of the ocean they found stratified layers of warm and cold water which protected them from depth bomb death. As above, so below.

This idea of limited power of flight on the part of the UFO of this earth interested me greatly, and I wondered if we physical humans were limited in what we might do as far as "space" travel is concerned. There is much bugling of achievement in the Soviet Union today, following upon their launching of the Sputniks. Boasts about space travel are being made, but if it is denied to others, perhaps it is also denied to us.

Question: How far above our surface may (the dark forces) penetrate, and can physical man penetrate this far? That is, will man ever be able, in physical form to penetrate this far?

Ashtar: "The limit of their altitude attainment is 125,000 miles. Physical man is also limited to this extension.. In the upper portions of it however, man in the physical form will exist only with extreme difficulty and after years of training and development. As previously described, when the moon's astral envelope or aura overlaps with that of the earth, commerce is possible. At times of contact, the astral entities from the core of Shan travel to the moon."

It is possible too, that the moon physicals visit us. Let us examine this last communication in the light of what we already know about such things. Consider the modern experimental pilot, thrusting his craft towards higher and higher Mach numbers. This intrepid man is clad in a special pressure suit which has the function of holding his body together and keeping his blood in the right portions of his anatomy. He must take along his own atmosphere to the altitude he is flying, and in some cases his airplane must be refrigerated to overcome the heat generated by atmospheric friction. If any of these things fail, the pilot meets "death." Why? Because, boiled down to essentials, the tolerances of his physical vehicle have been exceeded. Could it be that what actually happens is that the interpenetrating invisible bodies described by Dr. Cannon, leave the physical, or are *driven out* by conditions other than those of the earth's surface? It would seem that proper investigation of the behavior of these invisible bodies is the best hope for "space medicine," so called. The greater the altitude, the more acute this problem becomes, and as we get higher, we may find that many of the "laws" of physics, as we know them in our earth-bound state, are refuted or reversed, particularly the gravity-levity relationship.

A higher power has probably decreed where we shall or shall not go, and it looks as though eventually we may reach the moon in our endeavors. What might we expect to find on the moon?

Question: What is the nature of life on the moon? Are the moon people physical or astral?

Ashtar: "The moon people are physical in form and astral in allegiance. They are allied with Satan."

It is of course, likely that these moon physicals, if they be allied with Satan, would have access to the same powers that he presides over, including types of spacecraft. Therefore, we can add another faction to our UFO visitors, moon physicals in physical craft, which will probably appear predominantly around the perigee of the moon.

Let us remember also that song "By The Light Of The Silvery Moon" and keep our eyes peeled for silvery craft which could conceivably originate in the silvery, lunar sphere.

If things on the moon are as Ashtar states, it might be a wise policy to let the Russians get there first, at which time they will probably get what they deserve.

Nevertheless, let us not abandon the idea that we may already be the subject of visits by beings from our own moon of physical or near physical type.

Returning to consideration of the craft themselves, there have been frequent references in other writers' works to the Atlantean riddle. This same theme has recurred in this book through the explanations presented of the origins of certain of the entities present in the UFO. I now wondered if there was any connection between the craft from "downstairs" and the machines which flew in Atlantis, according to legend.

Question: Are these craft comparable to Atlantean machines?

Ashtar: "They are almost identical with those craft. They are, of course, made of a material that is akin to, but of a higher vibratory form than your own matter. They are, therefore, not normally visible to your optics."

The operative phrase here would seem to be "not normally," and this would seem to check out with the appearing and disappearing UFO which have baffled investigators for so long. Not normally visible but sometime visible, hence many of the observed UFO which disappear before the eyes.

Logically enough, the next question that arose was how to tell one from the other. Who was who and how can one tell? Can the separate manifestations be identified? I asked once more.

Question: Is there any broad general method by which the etheric or friendly craft can be distinguished from the astral machines from the center of the earth?

Ashtar: "As a general rule, you may conclude that all cigar-shaped craft are potentially hostile to your people. These are the craft from the center of the earth which have carried out and are carrying out hostile acts . . . . . Our craft are for the most part heel-shaped, or disc-shaped. This is a rule of thumb, as you term it, for distinguishing between them."

This information could be applied to many sightings, and perhaps some interesting and enlightening conclusions drawn. Ashtar here does not say, mark you, that ALL cigar-shaped craft are entirely and invariably hostile. He says "You may conclude." These are general rules, rules of thumb, and *not* ironclad formulae.

The cigar shaped craft have been observed starting fires and in other questionable acts, as described by Mr. Harold T. Wilkins in his two books.

Sightings involving clear *shapes* are seemingly in the minority while light manifestations are very abundant. Perhaps there might be a rule of thumb regarding light manifestations.

Question: Are there any basic rules, even if they are broad rules, by which the various craft can be identified when they appear solely as light manifestations?

Ashtar: "There is a broad general rule which may apply for the purpose of identification. It is not exact, but is a 'rule of thumb' as you call it. The true interplanetary craft, the ventlas of our forces will appear to your optics with a manifestation of colored lights, usually green, red and white. They will sometimes appear constantly red and green, other times they will appear to be flashing. Those of the satanic forces seldom exhibit color, but come with white or bluish white manifestation. This should aid you in selecting the ships with which you might have contact."

The ability of people to misinterpret these last two messages is unlimited. We are given a rule of thumb, and yet, people will persist in interpreting this as "you said *all* blue lights are *bad*." Things are just not that simple!

The proper course of action is to test the criteria, then accept them or discard them in accordance with what one learns from their use. I am still personally trying them out, and expect to do so for some time yet.

At least now something was at hand concerning the craft, their origins, general shapes, light emanations, and the matter of which they were constructed. The next step was to find out about the crews of these various UFO, and about those UFO that didn't seem to be craft.

In this field, surprises lay in store, and yet, over the months, much of what was offered from the invisible turned out to be true, and has been photographically documented as shown in this book.

All the UFO are not, and are not manned by, handsome cosmic gents who don't know the meaning of a dentist's drill and who possess hair of a length and beauty our earth ladies might envy.

CHAPTER FOUR

"GREEN MEN, MONSTERS AND SKEPTICS"

"The Water of the Air, between Heaven and Earth is of all things the Life."—Paracelsus.

The UFO picture was still far from complete, even though the communications were filling it in quickly and with bold strokes. Sometimes shocked, sometimes nonplussed, I nevertheless found much that was contained in these communications to be true. In many instances, explanations came to hand for observed phenomena, although these were not always direct answers, but rather signposts for the direction of one's reasoning power.

Some information had been offered about the various types of craft, now the question arose of the crews. What were the occupants of these craft like, if indeed they were all craft?

Referring specifically in this case to the machines purported to come from the center of the earth, I addressed the following question to Ashtar:

Question: By whom are these craft manned?

Ashtar: "They have a variety of beings in these craft. They may be human type entities in the astral body, in every way similar to yourself. They may be elementals, sub-human slaves of the astrals, or they may be astral monsters of a proportion and type likely to fill you with terror were you to contemplate them. In your measurement, many of them are gigantic, and horrible to behold. The craft may be manned by any or all of these creatures."

This statement threw new light on the many monster stories connected with the UFO. These monster stories have been grist to the mills of skeptical newspapermen, and stimulus to the diseased imaginations of Hollywood's writers of horror screenplays.

Because of the latter day plethora of horror films produced by Hollywood featuring a "space" angle, the general public has been led into that dangerous state of illusion characterized by the phrase "It can't happen here." In this respect, unscrupulous, money-grabbing profiteers amongst Hollywood's producers have rendered mankind a monumental disservice with their trash.

Is there any person interested in the UFO who has not heard the phrase uttered when he speaks of the UFO: "You've been seeing too many Hollywood horror films."? At this point, all serious discussion of these matters is drowned out by the gales of guffaws. And of course, this same attitude has been engendered amongst the gentlemen of the press, who also prefer to mock those things that they cannot understand.

The case of the Sutton, W. Va., monster story was investigated by men who are above suspicion, including some highly skeptical and case-hardened local sheriffs. In this Sutton story, now a matter of record and fully dealt with in other works, a craft was seen shooting overhead by a group of boys. Running up a hill in the direction the UFO had disappeared, the boys were confronted by a towering and horrible creature which gave off a nose-withering stench.

One of the boys switched on his flashlight and directed the beam at the entity. The monster promptly broke and ran or rather "bounced" across the fields, indicating that despite its size it was well-nigh weightless, OR NOT SUBJECT TO GRAVITY in the same way we are. Astral matter is not subject to gravity in the same way as physical matter, according to Ashtar. The creature could not bear *light* and it stank. Hollywood horror films? The only part they played in this incident was to make those who did *not* see the monstrous being impugn the observational powers of the boys who did see it. Their sanity also was impugned, at least tacitly.

The reader at this point is asked to look at some of the photographs in this book, taken by myself and my associate, of beings that were invisible to the eyes at the time they were photographed. Now, what would the reaction of the average person be if he were confronted by rudimentary beings of this type on the top of a lonely hill at night? The photographs I have taken certainly contain some instances of beings "horrible to behold" and little more need be said beyond the fact that science has a duty to the citizenry to investigate the nature of these beings.

In the Sutton case, it would appear that some strange freak of atmosphere, earth's magnetic field, vibratory level, climate, and a hundred other variables rendered astral matter visible to physical eyesight. This would account for the sighting of the craft and the visibility of the occupant.

It is obvious to the simplest intellect that the distaste for light exhibited by this monstrous being indicates its preference for the dark. Who said that there is no such thing as the "dark forces?"

Some of these high priests of pooh-bah who sit at newspaper desks would doubtless still prefer to call the whole happening an hallucination. From hallucination, there is the derivative term beloved of the press in cases where scores of people see the same inexplicable thing, the "mass hallucination."*

Another monster story came from Garson, Canada where a miner encounted a giant entity thirteen feet tall. The miner stated that this "spaceman" fixed him with a hypnotic stare. Contemplating the strange, burning eyes of this being, *and his six arms,* the miner collapsed in a faint. When he revived, UFO, "spaceman" and six arms had all gone.

In many a copy room, this incident was good for a laugh, but it happens to be classified by the Royal Canadian Air Force. One of the reasons it is classified is because the "monster" spoke to the miner, and while no official statement has been made about what was said, I have it on good authority that certain suggestions were made regarding a certain manufacturing plant by this "spaceman," to the miner.

In this instance, the technique is obvious. The "monster" portion of the story is released, and is immediatly blown down by all sources of public infor-

*Los Angeles Examiner November 7,1957

mation. The portion of the contact dealing with a threat to security, or with intelligence conveyed by the unsavory visitant, is hushed up neatly.

It is important to record here that this description again fits Ashtar's description of the form taken by some of these beings, "astral monsters of a type and proportion likely to fill you with terror were you to contemplate them." I submit that these descriptions fit very well, and even if newsmen were not filled with terror, I have no doubt that the miner, who was there, was justifiably terrified. Therein lies one difference between the closed minded skeptic and the open minded skeptic. The open minded skeptic gives the benefit of the doubt to the people having these shocking experiences.

The most dearly beloved of all chopping blocks in connection with the UFO are stories of the "little green men." Sightings of these creatures are strenuously pooh-poohed just because no qualified medical man has ever had one on the laboratory table at a recognized clinic (or have they?) Usually, the "green men" are passed off as drunken hallucinations or something equally convenient.

A well-known Los Angeles newspaperman, introducing a speaker at a saucer meeting in that city, once described those reporting these "green men" incidents as being the "weak voices" in the UFO picture. Apparently he didn't believe that there were such things himself, or he would have sought instead to fit these things into the mosaic, instead of taking the view that reporters of such incidents are "weak voices."

Can we dispose of the little green men so lightly and easily? I wish that it were so!

Little green men keep appearing, little newspapermen notwithstanding, and some of our people are having fights with them. The results, or rather, lack of results, have been amazing.

In 1955, a Hopkinsville, Kentucky farm family named Sutton had what can only be described as a pitched battle with a horde of these green men in and around their own farmhouse.* Like all good Kentuckians, the farmers had shotguns and plenty of shells. They fired off their full supply of shells, knocking the green men off the roof gutters and corners of the building, but they could not kill or injure them. Time and again shotguns were discharged at them from point blank range, with *no* effect. Does this behaviour not suggest interpenetrating matter?

Compare this incident now with the two related in Major Donald Keyhoe's "Flying Saucer Conspiracy", which occurred in South America. Many thousands of miles from Kentucky, these encounters have points of similarity which will not be obvious to orthodox science, but which have great meaning for students of the superphysical.

The first of the two encounters in South America took place on November 28, 1954, at 2 a. m. Two Venezuelans were on their way in their truck to Caracas for food. They saw a luminous sphere ten feet in diameter blocking a side street, and then a

---

*Los Angeles Times August 23,1955

hairy dwarf came towards them. One of the Venezuelans ran to a traffic inspector's office nearby, while the other one took a swipe at the dwarf with a knife. The knife simply glanced off the little man's shoulder as though it was *hard*. His body seemed equally hard and unyielding as the Venezuelan grappled with him. The arms of the Venezuelan were scratched and his clothing torn, as later verified.

A couple of weeks after this incident, two Venezuelan peasants spotted a similar round object hovering near a road. Four *little men* got down out of it, and *tried to drag one of the Venezuelans into the craft*. A fight ensued. One of the Venezuelans swung a shotgun on one of the dwarfs, only to find that the shotgun shattered on the creature without injuring or fazing him in the least. The encounter ended when the headlights of an oncoming car illuminated the scene. The three-foot-high assailants promptly clambered into their cosmic taxi and took off. This story was checked by the A. P. R. O. organization. *

Let us now examine these encounters for common points, and see what we can do to explain these events. The first point that does not seem to have registered fully with most investigators is that the creatures could not be injured by shotgun blasts, a knife, by grappling, or by a shotgun used as a club. They could not be damaged by any of these agencies which *we know* to be damaging because they are of an order of matter different from our own.

*Now Disbanded

In the Kentucky case, the shotgun blasts appeared to go right through the green men, which would tend to fit in with the astral theory. In the Venezuelan encounters, the beings were positively not astral, since our matter did not interpenetrate theirs. They appeared to be much denser, i. e. harder, thicker and heavier than the men they encountered.

There is an old Hermetic axiom which says "As above, so below." It seems to me that the heavier, denser, hairy and rudimentary dwarfs would logically have their origin at a point or level or density more subject to gravity than the surface of the earth. That is, they could originate below our surface.

In all instances however, the creatures were of a different matter than ourselves, whether denser or finer. They were, in addition, *hostile*.

In the second Venezuelan encounter, headlights caused the little men to depart in haste. Light dispels darkness.

There may be many interpretations of these incidents, but in my view, any interpretation which assumes these beings are of the same density as ourselves is bound to be incorrect.

We might note at this point, since it has a role to fulfill in later explanations of various phenomena, that the hairy dwarfs attempted the kidnapping of one Venezuelan.

Here is an additional discourse on this subject offered by Ashtar. It throws more light on the subject, if we choose to use the information offered.

Ashtar: "The silver spheres frequently reported, especially those associated with reports of little men as occupants, are a type of craft launched from the carriers of the dark ones. There are several facts which govern the sightings of these things. In the first place, a combination of atmospheric conditions and the physical condition of the viewer may render them visible. In the case of the monstrosities seen by only one person, this may be the case. Certain physical conditions in the viewer may render them visible to one man when they will not be seen by another man beside him. Hence the common term 'hallucination.' But the man who sees these monstrosities is seeing something very real, and while it can be said that the experience is a subjective one in a sense, it is also a view into the unseen worlds which surround and interpenetrate your own. Great confusion is caused upon your surface by the varying descriptions of little green men, little men in various types of clothing and so forth. Believe me, it would take many books to fully describe the many types of elementals who dwell in the invisible realms."

This at least gives us a new perspective on hallucinations. They may be *all* that they seem, in some cases.

In referring to this combination of physical conditions and atmosphere Ashtar is probably referring to those people who experience an extension of normal vision involuntarily. Modern psychiatry is lost in these matters of course, but the development of every individual consciousness is different. Consequently, what each individual is likely to see and interpret as real is also different.

The man with "tunnel vision" for example, may reach twenty-five years of age or more before he realizes that everyone else does not see the same way he does. The same thing occurs with persons clairvoyant from birth. Phoebe Payne, for example, the great English psychic and seer who has worked extensively with doctors on diagnosis, states in her book "The Psychic Sense" * that she was nineteen years old before she realized that not everyone saw what and as she did.

Any person who has studied superphysical things knows that starvation can bring on clairvoyance as the delicate membranes of the etheric "double" which normally screen out the sight of other worlds collapse through malnutrition of the physical body.

My own experiences in California's Mojave Desert are endorsement for the suggestion that location plays a role in the visiblity of these other worlds. The atmosphere in that area has a different quality than city atmosphere, aside from its cleanliness. It is highly charged with static electricity. The role that location has played in my own researches will be explained later on, but in my view, this Ashtar communication is valid.

As for the descent of various beings in beautiful robes in assorted sizes, offering different things to different people from different craft for different purposes, the best advice that can be offered to anyone

---

*"The Psychic Sense" by Phoebe Payne and Lawrence Bendit M.D.

dabbling with the UFO is *not to trust appearances*. In the astral world at least, changing appearance is a characteristic stressed in almost all writings on the subject, be they western or eastern in origin.

Returning to the monsters, unpalatable and all as they may be, there have been instances of pilots seeing some of these saucer entities and going out of their minds as a result.

One such incident, presented in what amounts to hearsay form in Major Keyhoe's "Flying Saucer Conspiracy," concerns an Air Force pilot in Hawaii. Sent up on a UFO scramble, he was alleged to have landed in babbling terror, after having seen the occupant of one of the UFO's.

I submit that if this pilot did in fact see the occupant his reaction would be understandable if we accept Ashtar's description of some of these beings. They are "horrible to behold".

The picture of the UFO which we have labelled the "Amoeba" also would be sufficient to scare any airman out of his wits, no matter how brave he might be or how well trained. Preparation for the impact of suddenly and without warning seeing an object as the "Amoeba", complete with what looks like a face, is not a part of an airman's training. I am extremely grateful that I did not see this object, but "saw" it only in the sense explained in the chapter on photography. Had I seen the UFO in question optically, I would quite probably have been in similar condition to the Pearl Harbor pilot. This reaction would have been intensified, if, as I now suspect, the object had absolutely no depth, i. e. was *two* dimensional.

The Pearl Harbor incident also harks back to Oct. 8, 1953, when a voice suddenly cut in on a Salt Lake City radio station and proclaimed : 'I speak from a space ship. You cannot reach me, but I can, with ease reach you. If you saw me, I should be so horrible in your eyes that you would be scared to death.'

This story was given a good dose of pooh-bah and put to bed. Now perhaps, with the photographs I have taken, and the incoherent testimony of a demented pilot to go on, defense officials and UFO fans alike will realize that perhaps, after all, the entity concerned was not kidding. For we cannot totally exclude from possession of telepathic ability, the strange beings that live in the sky illustrated in this book.

The finest sightings of little green men yet released to the public took place in the area of Steep Rock Lake, Ontario, Canada, a barren district whose life centers around the iron mines. On one occasion little green entities were seen hosing water into a hatch on their saucer, and then promptly departed when they became aware of several miners watching them. They have also been seen frequently in other countries near lakes, rivers and bodies of fresh water. Large quantities of water, involving millions of gallons in some cases have been found mysteriously missing following UFO sightings in various parts of the world.

Especially has this been true in England, where the tireless English writer and investigator, Mr. Harold T. Wilkins, has kept a tab on these activities. In some instances of water theft, a green scum (from the green men?) has been found on the water. This scum has defied laboratory analysis, and has disappeared shortly afterwards, in much the same way as the weird and misnamed "angel's hair."

In one Steep Rock sighting, the two observers noted that there was some kind of sweeping antenna, much like a radar antenna, whirling around atop the craft as it rested on the water. The purpose of this device was apparently to warn these green men in silver suits of the presence of living beings, for as the observers looked through a crack in the rocks, the antenna halted upon their very location. Instantly, the green men, who appeared to be somewhat transparent, ceased hosing water, vaulted into the hatch of the craft and took off. The people who saw them said that *they had no faces*. This should not be disturbing to UFO fans, who realize that our own Pentagon is full of faceless spokesmen!

There are two things that must be carefully reasoned in this case, for they are of paramount importance in my opinion if we are to solve even the outer fringe of the UFO mystery.

First, the green men in silver suits (remember again the "silvery" moon) were seen *stealing* water. This is unethical since it is theft, and must therefore in some measure characterize the beings who performed the act. It is probably the only single thing these UFO entities have been seen doing that makes any sense. It is an intelligent, if unethical action, the theft of water. What can it mean?

The viewpoint of a well-known UFO lecturer that they take water because we have plenty and would not miss it is just plain humbug. *Apologizing* for unethical acts on the part of UFO entities must not become a substitute for *analyzing* them. These entities, like all others, must be known by their fruits, not have their fruits deceptively packaged by earthlings who crave fame, glory and adulation.

These entities do not belong here, and therefore, have no more right to steal from this planet than any man has the right to enter a home where he does not belong and steal water or anything else. Let us beware of applying imported, extraterrestrial morality and forgetting our own.

Rudolph Steiner, the great anthroposophist, once stated that the entire solar system was built of primeval fog, which extended as far out into space as the planet Neptune. Fog is a form of water. Water is inseparable from *life*. Life cannot long endure without it. We cannot begin to fathom the extent of the esoteric knowledge to which giant intellects like Steiner may have been party. Nevertheless, we can use the writings and teachings of these exceptional beings who have lived among us to fathom what some of these unknown entities may be up to.

The UFO entities in this case did not want to be observed stealing water. They had taken special precautions to avoid detection. Their radar-type device apparently picked up the vibration or body force field of a deer which ventured near the edge of the lake. They were equipped to *avoid detection*.

They did not especially care about such things when they terrorized the Kentucky farm family, nor did the hairy men worry about being seen in Venezuela. In the Steep Rock Lake instance however, while at our frequency, they did not wish to be seen. Self-preservation is the strongest of all human motives, and it is not illogical to conclude that this theft of water is connected with *their very life.* Suppose that present astrophysical theories regarding the moon are correct, and that there is no water on its surface. Where would moon entities be most likely to get water in order to stay alive and keep their civilization functioning? From the only other body they could reach in their craft of course, the Earth!

I believe that the answer lies in part here, and in part is connected with the propulsion of their craft, something the Ashtar messages will aid us in theorizing about in due course.

Ashtar has already stated that blue and white lights, steady, are a possible indication of the presence of the craft of the "dark" forces.

Water is made up of hydrogen and oxygen, and hydrogen burns with a blue flame. Oxy-hydrogen flame is blue through white depending on the percentages of the mixture.

Objections immediately arise that "they are propelled magnetically." Is this so? Is it true that ALL are so propelled? My own experiences on California's Mojave Desert indicate that compasses are not always affected by these invisible craft. And the reader's attention is directed to the many instances of noise associated with the UFO mentioned by Mr. Harold Wilkins in his two UFO books. I refer to mechanical noise, to jet-like noise, and to those visual manifestations variously described as flame and exhaust at the rear and sides of the UFO.

These manifestations have been almost universally pressed into the molds of pet theory held by the various investigators. I submit that there is another aspect to UFO propulsion, relative to the method employed, which explains the speed limitations mentioned by Ashtar. The interplanetary craft of certain benign beings are propelled, in their terminology, by "polarized light," but this catalog of noise and fiery manifestations associated with some of the UFO and seemingly similar to our jets led to this question being asked.

Question: Are the dark ones craft propelled by the same propulsion system as yours?

Ashtar: "No. They use an electromagnetic drive on some of them. Others are propelled by a jet and hydraulic system. That their propulsion systems are less than ours is indicated by the fact that they have a maximum speed of 3500 mph in your measure. The different propulsion systems give rise to the different light manifestations, although these are a rule of thumb and not an arbitrary and fixed method of identifying the different types. They also have discoidal machines similar to our ventlas."

This communication is a classic example of the inability of a highly evolved and technical intelligence to convey to one untrained in these matters exact meanings. One cannot realistically discuss calculus with a janitor, and this represents the probable relationship between the mentalities of these etheric beings and those of ordinary people such as myself.

However, I do not think that I would flounder alone, but would probably find the finest scientific intelligences equally lost in these alien fields.

Analyzing this communication to the best of our ability, it would seem that the information fits in with what has been observed. It would appear that the "boys from downstairs" or their moon pals, or others whose habitat is the air around this globe, have come to our surface, and to our frequency, in order to steal water. They steal water to sustain their own lives, and also as a fuel for their craft.

As for the references to speed, we must remember that the many factions of entities involved probably are of many different types, and that general statements are going to be just that. If they are of an astral order of matter, then clearly all discussion of velocity is not within our province. Also, if they are of near physical nature, their speed attainment will depend on their ability to nullify gravity. In discussing space craft velocities, one must be extremely careful, lest he stumble unwittingly into one of the many traps with which the labyrinths of relativity abound.

However, we could tentatively assume, as part of our investigation, that certain of these hostile entities do operate under a positive velocity limitation. Later, we can discard this concept if our knowledge moves forward appropriately.

Nothing could be more in line with the character of unethical beings than that they should seek to delude us with tales of their fantastic powers, when all the time they may be entirely dependent upon our physical water for their feats in the air. To constantly impress on us, by means of gullible and untrained psychic channels, that their craft are magnetically propelled would throw our scientists and investigators right off the track! And their giant ships, purportedly miles in length, palmed off on the unwary as carrier craft, may in many cases be nothing more than enormous tanks for carrying water which they have stolen from our surface. Could it be that their known desire to avoid detection in this matter grows out of the fear of what might happen to them if their "water were shut off?"

Question: On occasions, elementals have been seen stealing water, or water has disappeared in large amounts following saucer sightings. Why is this?

Ashtar: "Water is a valued commodity in the center of the earth where they dwell. One of the reasons they come to your surface is to steal water, which they do from lakes, rivers, reservoirs and tanks as convenience dictates."

It is obvious from the tenor of this communication that Ashtar is leaving us to do a lot of our own reasoning and thinking. He is helping us to help ourselves, which would seem to be the sensible and ethical way. However, to this day I remain skeptical of the presence of astral forces within our earth. Skeptical, that is but *open minded.*

It might be said in support of this communication that if it is as hot in the center of the earth as both legend and theory dictate, water might be a "valued commodity." Who knows? Then there are those who hold that hell is as cold as ice! Perhaps the water is needed to keep the temperature down, if *this* is the case!

Certain helpful keys are being handed to us by unseen but benign beings. Without going off our rockers, or accepting all that is so offered blanket-fashion and without discrimination, let us try some of these keys. Let us beware that they do not fall from brash and dogmatically skeptical hands before they are at least tried!

There is, in my view, yet another use for water which these entities apparently desire so avidly. This is *concealment*. The condensation of a cloud of water vapor around a hovering craft is a means of concealment that I have personally seen utilized by some of the UFO. I have also photographed this phenomena. One photograph of a cloud which I *believed* contained a huge craft appears in the illustrations. I have others showing the well-known bell-shaped craft concealed in this way.

I first stumbled into the concealment aspects of the UFO at the Spacecraft Convention held at Giant Rock, California in 1956. My attention became drawn to a section of the sky beyond the head of Mr. Frank Scully, who was speaking at the time. I saw several clouds appear and disappear in a strange fashion, for they hung motionless and unaffected by the wind that blew quite strongly. I noticed that the vapor trail of a high flying plane was rapidly dispersed, while these clouds continued to materialize and dematerialize.

I took off on foot across the desert for perhaps a couple of miles, until I came to a point on the desert floor just below the clouds. As I watched, I saw one of the clouds suddenly begin to thin out and disappear. As it did so, I perceived a reddish glow with a rotary effect within it for a few brief seconds. Then about a minute later, I perceived the reddish glow once more, but it was obscured almost instantly by a condensation of vapor which gave the appearance of a small cloud.

These happenings were repeated time and again, and after I returned to the convention site, I took another bearing through the rocks and found that the clouds had *not* moved.

From this series of events I deduced that as the craft became visible, for some reason or another, it was capable of condensing water vapor around itself, remaining invisible to the eyes, even though it was probably at our visual frequency, inside the cloud.

I have not the slightest idea as to the nature of this craft, if indeed it was a craft, but I was convinced that it had the power to conceal itself by creating clouds or cloudlets.

Clouds require water from some source or another, and we know that water is being stolen from our surface by some beings who do not wish to be seen doing it. There is likely to be a link between these things. The desire for water, the use of water as a fuel or propellant, or concealment agency, and for the support of life. To these things we may add other services and requirements unknown and probably incomprehensible to us.

Summing up the evidences of the senses and the supplementary information offered from invisible friends, we find that we have a fairly good picture of who's who in spacecraft. The satanic or hostile craft, earthbound and possibly based in the core of our own planet, and on the moon, are likely to give off a bluish-white or white or blue emanation. They are believed to be in some instances, cigar shaped, with the light manifestations tending to be constant rather than flashing.

Benign craft of an etheric nature believed to be beneficently directed, are believed to give off red, white and green emanations, which are flashing rather than constant.

The satanic craft are understood to have a maximum air speed of 3500 mph although their ability to overcome gravity may give them much greater velocities. This remains to be elicited.

The etheric craft, being solely mind constructs of the fine matter of the etheric level of life, travel in accordance with thought power, and therefore velocity as such has little meaning for them. Within the earth's magnetic field, they probably travel at the speed of light, whatever that is.

Of course, the intelligent investigator always wants to know about *proof!* Predominantly here, I am advancing *beliefs,* to try and erect a tentative structure which we may elaborate on later as evidence accumulates. We are not entirely without proof for these criteria handed us by a benign invisible.

Both Major Keyhoe and Harold T. Wilkins have mentioned this particular sighting, which would seem to be of one of these satanic machines described above. In "Flying Saucer Conspiracy," Major Keyhoe relates:* "In Munster, Germany, a movie projectionist named Franz Hoge reported watching a saucer land in a field. Hoge, said I.N.S. (International News Service), discovered a cigar-shaped machine hovering six feet above the ground, giving off a brilliant blue radiance which nearly blinded him. Just after this, he sighted four small, peculiarly shaped creatures with thick-set bodies, oversized heads and delicate legs!"

This would, in terms of our information from Ashtar, make this a machine from the center of the earth, or possibly the moon, manned by elementals. It fits exactly by shape, light manifestation and occupants, the descriptions offered by Ashtar.

To the mortification of the high priests of poohbah, Major Keyhoe goes on to state: "Because of Hoge's story, some UFO censors found it easy to ridicule the entire I.N.S. account. But a report, flashed to them from London on October 14, hit with harder impact. For this was an officially confirmed Royal Air Force report."

The Royal Air Force is not exactly a knee-pants and hoop-rolling organization, and confirmed reports by RAF intelligence are likely to have been given the most rigorous scrutiny by men skilled in the science of defending their country.

The well-known UFO lecturer in California who gets titters from his female audiences by describing these handsome and magnificent spacemen has not yet been known to suggest that they also take into

*"Flying Saucer Conspiracy" page 207. Henry Holt and Co.

account that some UFO beings have thick-set bodies, oversized heads, delicate legs and stand only four feet tall. The chances of cosmic romance between lovely ladies of earth and some of these weird visitants would appear to be remote.

There is a massive catalog of these unsavory manifestations connected with the UFO, which devout supporters of the saucer phenomenon are reluctant to explain and which are difficult to explain as long as one holds to the assinine and illogical viewpoint that all who come from elsewhere in the universe must perforce be beautiful, loving and advanced in morals and ethics.

These unsettling phenomena cease to be unsettling when one realizes that we have a multiple manifestation among the UFO. The moment the duality of life as we know it is also assigned to the UFO, the pieces in the puzzle begin to fit. Just as there are numerous nations on earth with varying morals and ethics, so are there probably orders of civilizations throughout the universe that vary as much or more in this respect as our own puny globe.

Some religions teach that life is eternal, and happening to believe that this is so because I find any other viewpoint untenable, I also find that this helps me to comprehend a great number of the strange elements present in the UFO field. If life is eternal, then clearly there must be a place, or a plane, or a vibration to which we adjourn when we leave the physical plane of life.

I do not for a minute think that all the Hitlers, Mussolinis, Rasputins, Stalins and other tyrants of history are just permitted to wander around the universe willy-nilly. I think that for all these once humanly incarnate monsters there must be an invisible gathering ground, and that it is likely to be in the "astral workshop" mentioned earlier.

A further communication along this line reached me from the pen of a famous English writer who reviewed some of the Ashtar messages and whose opinions I value deeply. Without coming into contact with those portions of this book which deal with the origins of Satan and his followers, but referring directly to some monster sightings, this man wrote to me regarding the statues at the Portico of the British Museum, as follows: "Have you ever studied, *close up,* the brutal giant men, with extremely forbidding features, jutting jaws, thin lips, stern and merciless eyes, sadistic in the extreme, brutes with brains but not our kind... I mean the giants of old Mu?"

If life is eternal, is it not possible that these beings went somewhere when they died.? Is it not possible that they still exist in the unseen worlds about us, serving the same master now that they served in old Mu?

When we deal with these matters, I believe we are coming close to the real reason why the governments are loath to reveal to the general public of the world what they have deduced. The concept of astral beings, and different orders of matter may still be beyond any rational acceptance by governments and their satellite scientists. Even if such concepts were accepted within the governments, could they be imparted to the nations through presidential fireside chats and parliamentary bombast? I think not!

Imagine for one thing the reaction of the churches to government announcements along the line of these communications. Teaching as they do that there is no death and there are many forms of matter these communications would not find acceptance among contemporary theologians whose vested interest lies not so much in the truth, but in perpetuating their particular sect or creed and in the weekly sermon which is little more than pure opinion.

However, I believe that there may be some highly placed men of signal spiritual quality who already know what faces humanity, and who might well be the agency through which this information will be passed on to the nations.

Supporting this viewpoint is the report that General Douglas MacAthur had made a remarkable statement to Mayor Achille Lauro of Naples, Italy during a visit to New York by the Italian dignitary. On October 8, 1955, Mayor Lauro stated to the press: "he (MacArthur) believes that because of the development of science, all the countries on earth will have to unite to survive, and to make a common front against attack by people from other planets."

General MacArthur is regarded by many people as a reactionary, but even the bitterest of his detractors can hardly ignore his latter day efforts to warn the world of the consequences of further wars.

Failure to listen to his warnings issued on the deck of the "Missouri" in Tokyo Bay in 1945 brought the world into the horror of Korea and to its present unstable and immoral condition.

To me, General MacArthur is one of the noblest figures of our age. He is one of the most splendid Americans of all time, and it may well be that his deeply spiritual approach to life caused him in this case to catch a glimpse of what confronts the human race. He has been right before!

At the very least, this incident with Achille Lauro indicates quite clearly that one of America's greatest men views the matter of interplanetary flight with deep concern, and not as a delightful adventure portending nothing but limitless pleasures.

CHAPTER FIVE

SCIENCE — THE INFALLIBLE PHALANX

"We need to disenthrall ourselves."
— Abraham Lincoln

In the book "Into This World and Out Again" by George Van Tassel, the statement is made that there are three sub-stations established in the "vortex of the Earth," whatever that might be.

One of these stations, named ShanChea, is purported to be fifteen hundred miles square in our measure. Yet, we cannot see this enormous object! Because we cannot *see* the object, the scientific attitude has been that such statements are pure balderdash and properly belong in the science fiction field.

The information regarding these satellites was on record before Dr. Tombaugh and Dr. LaPaz located the so-called moonlets, and may be the reason that these eminent gentlemen began looking for these objects. Until their location, astronomers were quite sure that there was no heavenly body closer to the earth than the moon. But even Dr. LaPaz and Dr. Tombaugh could not "see" the moonlets, which were alleged to have been detected by radar.

Major Keyhoe is a careful reporter, and in "The Flying Saucer Conspiracy" he relates the confusion and upset caused at the Pentagon by the discovery of these orbiting bodies. This information was not given to Major Keyhoe, but slipped out by mistake when an old service friend of Keyhoe's saw him emerging from Admiral Radford's office. Major Keyhoe's friend automatically assumed that Admiral Radford then Chairman of the Joint Chiefs of Staff, had confided the secret to Keyhoe, America's leading saucerer. This officer friend of Keyhoe's let the cat out of the bag, then quickly tried to cover up on realizing his mistake, but it was too late. The story is well-known through Major Keyhoe's book.

If these objects are "moonlets", as stated, the point immediately arises of the Pentagon's sudden top-level interest in meteoritic bodies and the science of astronomy. If the objects are "moonlets", why is the whole subject a hush-hush one within the walls of the five sided giant?

The answer lies of course in the fact that more is known, and more is theorized about in these matters than the military are prepared or willing or authorized to tell the people.

It is well to remember that in the United States, the military does what the President orders, and that the reverse is not the case as some people seem to think.

The claim has already been made that the reason these "moonlets" picked up by radar, cannot be seen with the ordinary eyesight, or with telescopes is because they are coated with some kind of light reflecting paint.

If these objects are the size claimed by Van Tassel and my own communications endorse his in this respect, then they would easily be visible to any powerful telescope. After all, an open backyard and good binoculars permit us to view "Sputniks" and the "Explorer." Anything large enough to send a radar echo back from five to seven hundred miles out in space should be optically visible, unless there is a *modifying factor*.

It is here that science in the orthodox sense breaks down, for such an object would have to be visible by all orthodox astronomical reckoning. It is when we turn to sections of superphysical teachings, and to elements of some of these communications offered by invisible beings that we begin to make sense out of happenings like this.

All communications from unseen beings are threaded through with references to polarity which have no meaning to our science. After all, orthodox science does not even yet admit that gravity has an opposite and equal force, named levity by superscientists of the Goethean school.

According to George Van Tassel, these space stations are positive to light, and therefore not visible to our positive polarity vision. Which sounds more likely to be fruitful in discovering the truth — a new theory on light polarity, or that of light reflecting paint? The latter would seem to be no more than a subterfuge while astronomers go back to their drawing boards!

I cannot personally say whether or not this theory of light polarity is right, because I do not *know*. I do *know* and have photographically demonstrated, that living beings do exist in a vibratory level that our eyesight cannot respond to, and that there is evidence, that such matter interpenetrates ours. It even appears in some instances to be two dimensional, and the objects unquestionably fly. I have one photograph which would not reproduce suitably showing four of these "bladders" in flight, taken with infrared film over the Mojave Desert.

If one accepts the idea of light reflecting paint, one is also forced to accept the existence of some strange cosmic painter, spraying or brushing on his paint five hundred miles out in space.

Question to Ashtar: What can you tell me about ShanChea? How long has it been orbiting the earth? What is its size relative to Schare, the other base you have near us?

Ashtar: "ShanChea is the earth-child satellite, and has been orbiting your earth now for almost two thousand years since the appearance of Jesus the Master upon your surface. Fifteen hundred miles square it is a complex assemblage of instrumentation which permits constant surveillance of your surface and the beings upon it. Before very many months, have passed, although we will at this time give you no earthly measure of when, ShanChea will pass through the atmosphere of Shan, and a great commotion will be caused by its appearance. It will be visible to physical eyes at that time. Our present altitude is five hundred miles and varies considerably from time to time. Schare is different to ShanChea in that it is a Quadra Station. It is not square, but what you call spheroid. Its purposes are also different in our system."

It is essential to note that practically all this information was already in my mind from other sources at the time the question was asked, and that this information is neither new, nor necessarily valid. We cannot possibly just assume a connection between Jesus and this alleged space station, for example. We should have more substantial proof than a vague phrase like this.

Also, no mention is made of any other base besides the ones specifically mentioned by me in the question. For example, do the "amoebas" or "bladders" also have a space base from whence they hail? And is there not a possibility of other worlds of an order of matter closely akin to ours near us in space, just as invisible as the so-called "moonlets?"

The return of radar echoes indicates that these objects are real. But they are not visible to the human eye. Therefore, we must revise our concepts of reality to embrace facts of this nature. We must devise new theories, and test them out and not hesitate to jettison them when they do not fit the evidence. And we must rip from the shoulders of orthodox science the mantle of infallibility which has become so patently ill-fitting in recent years.

As we enter this spectacular air and space age, powerful currents are washing against the structure of orthodox science. Feeling their ivory tower teetering, astronomers, physicists and astrophysicists, to name just a few, doggedly hang on in their uncomfortable world, instead of stabilizing and stimulating scientific thought by making their peace with the superscientists.

Superscientists are not supermen. Nor do they deal with the supernatural. In fact, there is no such thing as the supernatural for all is a part of Nature. Superscientists are men and women who seek to become deeper readers of Nature's book who seek to understand causes instead of effects. They are not persons who think of themselves as demi-gods for the matters in which they deal have the power of rendering vanity the most assinine personal quality in existence.

From the day that the orthodox scientific fraternity joins hands with the superscientists, mankind's progress will make gargantuan strides. Meantime, however, most of them prefer to lock horns with the superscientists, whose door is not barred. On the contrary, the orthodox scientist will find that a warm welcome awaits him, so that both may go forward together.

Under the panoply of what is termed professional ethics, mighty crimes have been carried on through the ages against mankind and civilization. The most energetic, successful and well-directed scientific work has always been in the perfection of destruction. It is so today throughout the world. In America as of the date of writing, over $400 million has already been expended on the development of the "Thor" and "Jupiter" missiles. Meanwhile, millions die annually in this world from cancer, tuberculosis and other killer diseases.

Despite this sombre state of the world, scientists gathered recently in San Diego, California for the First Astronautical Congress held in America, were discussing the possibilities, probabilities and plans for the first war in space!

It is because such processes revolt the intelligent man that science is steadily falling from grace in the eyes of the thinkers. The destructive minds in our world's scientific fraternity have gained ascendancy over the constructive minds, and as the gasping taxpayers pour out rivers of treasure, greedy lunatics wait to squander it on bigger and better destructive devices.

The earth today is being thrust into a new age by evolutionary forces which transcend the power of our science, both in its ability to understand and to meet that age. It is an age wherein a cluttered and dogmatic mentality will mean intellectual strangulation. Yet dogma is one of the granite plinths of our orthodox science.

Consider this question of habitation within our own earth for example. It is very easy for theories to be fabricated, in fact it is necessary to *invent* them to cover what is within the earth. It is therefore easy for science to say that nothing could live in the center of the earth, perhaps nothing like *us* could, assuming of course that the center of the earth is what we theorize it is.

Scientists do not *know* what is in the center of the earth, and present day theory can hardly even be termed an educated *belief*. You cannot believe something once you know it. I know, and have shown that strange creatures fly in our atmosphere which are invisible. I do not know whether they could live in the center of the earth or not. But what right has science got to say that *nothing* could live there? It has only the right that it has arrogated to itself and the evidence would indicate that many of these assumed rights are pure fraud against the uneducated people of the world who look to science to lead humanity and not obstruct progress.

If my researches prove nothing else, and do nothing else for humanity than to prove that conditions for life as ordained by orthodox scientific belief are positively *not universal*, my time and effort will be well spent. What the conditions of life are for other orders of beings, be they physical, astral or etheric, I do not know, but in this ignorance I am but one of an enormous mob of ignoramuses whose membership includes the best medical brains on earth.

Since the entire chain of events and evidence brought forward by this book has been forged by the Ashtar communications, I feel that there can be no objection to tentative acceptance of the seemingly valid information that is offered. Time will clarify it all.

Astronomers, who seem to be less cocky in recent years, are gradually coming around to making admissions of various kinds regarding life on other planets. Those who make concrete statements in this vein are still looked upon as being radicals within their own profession, and I know at least one eminent astronomer who has been so shaken by the evidence he has encountered regarding the UFO that he forbids all discussion of it in his public appearances. A few years ago he wrote precipitate comments on the subject which today reveal his then arrogant ignorance of the subject.

As he is manfully struggling to get himself off the hook, I shall mercifully not mention his name, but he knows who he is, and he is probably only one of dozens.

Dr. Harlow Shapley of Harvard University was one of the first leading astronomers to try and drag his science into phase with the march of events. He created a stir a few years back when he said, or wrote "we are not alone." His subsequent articles have shown a broadening viewpoint and a progressive attitude.

Then there are others who make what appear to be dogmatic statements about the UFO and their points of origin. In one very strange instance, we have Sir Harold Spencer Jones, Astronomer Royal, who must know more about flying saucers than revealed in the statement he makes to the effect that:

"I can say quite definitely and with absolute assurance that none of the flying saucers have come from another planet."

We are going to need a mighty bushel to hide this gentleman's light! But let us examine his statement carefully. Sir Harold says "quite definitely and with absolute assurance," which would mean that he must have proof, since he is a scientist. Without proof, this kind of talk from a scientist will stick in the most gullible gullet!

Now his second statement is interesting, that "none of the flying saucers have come from another planet" in this respect, Sir Harold *may* be correct. If they are, in fact, inhabitants of this earth's own envelope, and are of this earth even if not physical, Sir Harold could be right. I am perfectly willing to admit that this is possible.

If, however, Sir Harold is referring to experimental rockets and other craft of British, American or Soviet manufacture, then he is unquestionably wrong. These undoubtedly explain some sightings, but do they, will they explain the "Amoeba" and will they explain the two strange invisible faces I have labelled "The Peekers"? I think not.

Astronomy is an observational science. It is based on the physical eyesight, which is a very limited faculty. The eye registers only an infinitesimal portion of our own spectrum, and we cannot see anything outside that narrow band of wavelengths without instruments or special film emulsions and filters. The basic dogma of astronomy is based on the physical eyesight, which, unaided by telescope, clearly informs the mind that portions of the heavens are blank. The instrument added to the eye to *intensify* its power gives the consciousness a completely new picture, and a new concept of reality.

We shall find in due course, that *extension* of the vision also presents new concepts of reality, and in the meantime, we should go slow and soft in accepting astronomical dogma as being exhaustive.

In general, these invisible beings seem to have a certain amount of contempt for much of our scientific "achievement." They could be looking on us with the same benign indulgence that we would reserve for the Congo native who first begins to master the operation of a simple piece of machinery.

With reference to astronomy, Ashtar has this to say: "The most outstanding fact about your astronomers is their unwillingness to learn. There are two basic things in your astronomy which lead to most of its erroneous concepts. Light is not understood by your technology, and light-years, the derivative measure of distance, which is based on a misunderstanding of light itself. On Shan (Earth) you have not yet begun to understand the true nature of light. We do things with light and through light that would astound your most brilliant minds. These erroneous concepts held by your astronomers lead to treacherous errors in the theories which are based upon them. These continue to compound themselves until the whole fabric of astronomical theory on Shan is stretched upon an entirely false framework. Errors of this nature are doubly compounded with the use of the spectroscope for viewing other planets and deciding by this means whether or not life is possible upon their surfaces. To begin with, all conceptions currently held by your astronomers regarding the nature of atmospheric gases, temperature and pressure upon other planets are incorrect. My own home planet, Venus, for example, has many elements in its atmosphere which are unknown upon Shan. Added to this, your scientists do not know the composition of their own atmospheric belt, through two hundred and fifty miles of which they are looking at the light from other planets. And then, space itself is something of which they know little, although their estimates of what it is are quite authoritative in their eyes. They imagine it to be a vacuum of some kind, whereas in fact it is filled with a host of things, including etheric currents of which they know nothing. The spectroscope cannot even perform with remote accuracy under these conditions, especially as the outer limits of your own atmosphere are extremely variable and uncertain as to content."

My own photographs indicate that space is filled with a host of things, without a doubt. Many leading thinkers also discard light-years as nonsensical. The speed of light within our own magnetic field and speed of light (if indeed it travels at all) in "free space" may be unrelated. Ernst Lehrs, Ph.D., in his formidable book "Man Or Matter" touches on this same subject with an approach that is guaranteed to stimulate the thinking power of the reader. In my view, since we have never measured the speed of light in "free" space, whatever that is, it is unscientific to invent the light year and attempt to force the structure and behavior of the universe into the rickety mold we have thus fashioned for it.

It seems to me that re-examination of our astronomical concepts with a view to *overcoming* the light year could revolutionize contemporary thought, and open broad vistas for mankind.

If able, trained and undogmatic men could have access to the information possessed by more advanced beings like Ashtar, a great scientific harvest awaits. Clearly, an untrained person like myself can only be presented with the most rudimentary ideas, such as those mentioned above.

Unfortunately, professional vanity is affronted by the suggestion that experts are only relative to the standard of knowledge currently possessed by mankind, and this in no way equates with all that there is to know about the universe and how it functions. Last century's experts would be today's dunces.

A splendid example of an astronomer unwittingly endorsing a "spaceman's" statement is found in the remarks made by Professor Norman Berrill of McGill University. Professor Berrill alleges that human beings our size and shape are found only on this planet and never on any other planet from here to eternity. This amazing statement from a scientist, as baseless as it is bald, unites ridiculous assumption with the truth about some of our interplanetary visitants *according to themselves!*

Berrill speaks of never on any other planet from here to eternity, and the intelligent reader is therefore bound to conclude that this illustrious gent has uncovered the limits of the universe. I would like to see a treatise on its limits, which would probably resemble ancient writings about the edge of the world in the days before Columbus.

Then there is a question of evidence. Observational evidence must exist for such a statement from an eminent member of a profession based on observation. Or must it exist only when the "other fellow" propounds a theory?

Professor Berrill's statement also points the way to the true nature of these interplanetary guests of ours, as *they* tell it. Ashtar describes himself as an etheric being, that is, his body, or vehicle of consciousness is made of a more tenuous substance than ours, which vibrates at a rate physical vision cannot encompass. Hence, they are invisible to us. Also, they are not subject to gravity in the same way.

Professor Berrill says human beings our size and shape are found nowhere else. Perhaps the etheric body will thrive in an atmosphere of a chemical composition lethal to the physical form, if indeed, it needs atmosphere at all. If what Professor Berrill says is true, and it is too bad he cannot prove it, it is support for the etheric concept of life on other planets in this solar system.

This entire reasoning process also lends further credibility to the theory that the UFO are predominantly invisible. That is, to our limited and feeble eyesight they are invisible, but they are none the less real for that, as the "Amoeba" and "The Peekers" indicate. We must free ourselves from the stupid and baseless dogma that physical eyesight and reality are always and essentially interconnected. Any optical illusion will prove that this is not so, and it appears to be also true of worlds and beings which vibrate at a rate imperceptible to our vision.

There is *observational* evidence to prove that the etheric concept has validity, strange as it may seem. The classic case of this is the New Zealand sighting where a large fusiform "spacecraft" as big or bigger than a DC6 airliner exploded over that country's Southern Alps leaving a tall column of smoke.

If this physical theory is to hold true for *all* the UFO, there simply had to be wreckage of this large craft. None was ever found. If a Constellation or DC6 blew up in this manner, the evidence would be irrefutable of such an event. Here again, orthodox science is bankrupt. Can we get an explanation of these happenings from an invisible intelligence? We can, and it is once again up to the individual to accept or reject as he sees fit. It is Ashtar who speaks: "On occasions, craft crash for one reason or another, not the least of which is the striking of giant nodes in the space currents on which we travel. These disintegrate the craft, even as a surge of power will burst open an electric motor not protected by fuses. Sometimes, in the disintegration process, the etheric matter may be rendered visible as it becomes subject to gravity and lands upon the earth. It will, however, rapidly degenerate into jelly, then into liquid, and then into gas, from which condition it returns to the universe once more."

Examples are legion of jellied matter falling from the skies through the ages. Evaporation always seems to have taken place. Could it be that some of these strange falls of jellied substance are the remains of crashed spacecraft which ultimately return to their true, or etheric state? The Ashtar message, plus the known facts, produces a logical answer. There is no reason why it cannot be tentatively placed in the mosaic.

A feature of our modern civilization is awe of the scientist. The mass of the people, untrained and comparatively uneducated adopt the attitude that "the scientist should know" and therefore automatically accept the scientist as an oracle, whether the scientist likes it or not, or seeks it or not.

In the past decade especially, as we contemplate the wonders science has wrought which improve our way of life, we tend to become slack jawed and pop-eyed over the extent and depth of present-day scientific knowledge. Everything from rockets to the stall shower has been scientifically treated, until finally we are giving the atom, the building block of the universe, the scientific treatment.

This particular activity has been the subject of persistent comment by various channels of communication with unseen beings. There are thousands of communications being received all over the world by mediums, psychics and other channels which all dwell persistently on this atomic bomb and its more prodigious descendant, the H-Bomb.

We have already noted in one communication presented in this book that some of the fireball manifestations are purported to be "nullifiers" for radioactive effects which we do not know we are releasing. Be this as it may, and we have no proof either way, other communications hammer on the "atomic meddling" theme to the point of suspicion.

These messages dealing with radioactive effects and atomic blasting need to be considered with extreme discrimination, for they are dealing with a field in which our best minds are still not at home. We have very little proof to back up any of these communications and must be careful. But we must also use extreme discrimination in regard to what our own scientific men tell us about these things. They speak in such pontifical terms in some cases, that the public is often led to the conclusion that they always know what they are talking about.

Particularly is this true of the nonsense dispensed by our scientists about the weather of the world not being affected by the atomic bomb. Meteorologists have stated that the weather system of the planet is too big to be affected by such a relatively small blast as a hydrogen bomb.

The evidence, which they presumably examine before pontificating, is clearly to the contrary. Consider, for example, this newspaper report of June 22 from the United Press, dealing with this particular subject. It appeared in the Los Angeles Herald-Express of that date: "Brisbane, Australia: Rain was reported falling today in the mining center of Kuridala, 1650 miles east of the Montebello Islands where Britain exploded an atomic bomb Tuesday.

"The report followed announcement Wednesday that an atomic cloud from Britain's test atomic explosion in the islands off Western Australia had drifted eastward over the mainland. Residents of Kuridala were warned not to drink any of the rainwater. A Queensland University atomic scientist said it appeared that the experts who exploded the nuclear test device last Tuesday had failed to get correct weather reports. Government officials and physicists had emphasized that the cloud posed no danger to human life. Today, Queensland prospector Jack Tunney said he detected the "hot" rain and recorded radioactive counts up to 2,000. Queensland University Professor Webster commented 'If any of our laboratories developed radioactivity as high as that we would be quite concerned.' "

In the rudimentary schooling given to most people, rain, clouds and wind are part of the weather. Meteorologists may not consider that these are part of the weather. I do regard rain, clouds and wind as part of the planetary weather, and since these were the agencies through which radioactive rain was delivered to a site of human habitation, I cannot accept any meteorologist's statement that atom bombs do not affect the weather.

The physicists stressed that there was no danger to human life, but said, don't drink the rainwater. Was the rainwater dangerous? If it wasn't dangerous, why forbid people to drink it? This particular incident indicates the kind of haggling and boggling scientists can exhibit over things where their knowledge is incomplete. They are not infallible.

What other effects are wrought by the release of radioactive particles we cannot tell. Perhaps these particles, after ascending to great altitudes, are affected by the earth's polarity to produce strange weather effects. The following summation of atomic blasting effects was offered by Ashtar: "I have information for you concerning the testing of hydrogen derivative weapons at Eniwetok. The continued release of these blast forces, both at Eniwetok and elsewhere on Shan has produced a change in your planetary weather system that is irrevocable. Your seasons, as you term them, are now moved forward eight weeks over past records. Thus, summer extends into October, winter into the end of April. These changes of course, are permanent. In addition, the axis of the planet has been disturbed producing irregularities in astronomical observations which are already concerning your scientists. We find it the height of fallacy and folly that your meteorological men cannot correlate these changes to blast forces released. The blasts penetrate not only to the limit of your atmosphere, but beyond into the solar system of which Shan is a part, and into other solar systems. Reactions are produced by this fission of which your scientists are presently unaware. These reactions affect the atmospheric envelope of Shan, through which they are conveyed first by the power of the explosion and subsequently by the planetary wind system. These are not predictions, my friend. These things are in effect today."

There would appear to be some elements in this communication that are incorrect in the light of the present knowledge, and others which may be correct in the light of what we don't know.

Certainly the weather was knocked out of kilter up to the summer of 1957, but the early advent of the 1957-1958 winter would seem to be at variance with the concept advanced by this being that the seasons are moved forward.

The ferocious battering administered by Mother Nature in the May and June of 1957 to the Texas-Oklahoma-Arkansas-Indiana areas would appear to be little short of extraordinary in these days of flood control and civil distaster organization.

The following communication came immediately following the much publicized and much delayed air-drop blast at Eniwetok, in the 1956 series of tests:

Ashtar: "Scientists at Eniwetok have exploded yet another fission device. You have already experienced the drop in temperature. There will be more. There will be considerable climatic havoc on your surface. The effect of fission upon the weather is obvious to everyone except the scientists, who do not *wish* to believe what is obvious to the simplest entity. Your scientists are not satisfied with the sun they have been given, but prefer to make their own. The Creator's work is not good enough for them. This prospect fills us with sorrow."

The tone of this communication is one of resignation, as though the personality concerned could hardly bear the comtemplation of what was happening. With it all, there was the strange atmosphere of inevitability.

Was there, in fact, "considerable climatic havoc" upon our surface following these blasts? With only U.S. newspaper reports to go by, I submit that there was. Immediately following the May 21, 1956 air-drop bomb, the city of Indianapolis was inundated by torrential rain. This flood was unprecedented in the area. The Idaho floods suddenly worsened, as hundreds were driven from their homes. An undersea eruption took place off the Hawaiian Islands, and record-breaking low temperatures, plus frost, inflicted at least five million dollars worth of damage on farmers in U.S. northeastern states. Many northern New England towns had temperatures in the 20's at the *end of May*. European records of weather at the time are also far from pleasant.

The communications of alleged spacemen, or unseen intelligences, may not always be right or acceptable by our standards, but the same criteria apply to statements by our own scientists who also make massive blunders. What the final verdict will be on the atomic bomb we cannot even conjecture.

Communications relative to the atomic and H-bombs fall into two broad categories, in my opinion. Communications which deplore the practice and use of the weapons, and yet seem to look on it as inevitable, and communications which are of a more frenzied nature usually urging us to "write the President" and so forth.

There being not the slightest doubt in my mind that there are *at least* two factions involved in the UFO, it was a logical conclusion that two factions would have varying viewpoints on this matter, just as they seem to have on other subjects. It seemed to me that some communications through other receptors, that I felt were originating in the malevolent camp revealed these entities to be a whole lot hotter under the collar regarding the Bomb than the benign beings. Some of these communications from the malevolent side were put together so cunningly that they would delude all but the most alert.

I began to reason that perhaps this fantastic blast force we had uncovered was as lethal to them as it is to us, or perhaps *more lethal* if they are made of finer matter than we are and *cannot leave the environs of this earth at will*. Perhaps the Elemental Spirits of of the Air, spoken about by the Ancients, did not take kindly to these detonations.

Believing that mankind gets nothing into his hands that he is not intended to have by the Almighty, whose purposes remain inscrutable to us humans, it followed that perhaps there was another purpose for these bombs beyond the one they presently have. This present purpose consists of Uncle Sam waving his bombs across the gulf that separates him from The Bear, who waves his in reply. In other words, has the Almighty, functioning on a plane and on a scale beyond the most extended human conception, delivered to mankind the weapon he needs to offset and check these satanic forces?

Let us examine this concept a little more closely. If we envision the Almighty as a God of Love, it is illogical that he would leave all his offspring to be simple, defenseless victims of Satan. Is it not logical that instead he would give us some physical protection against these forces, since we have for the most part rejected spiritual aid?

After some months of meditation upon this state of affairs, I resolved to ask my unseen communicator:

Question: It seems to me that there is too much unanimity in the objections received from the unseen worlds about atomic blasting. I also note that you seem to regard a certain amount of destruction as being part of our present lives. Is it therefore true that atomic bombs destroy in some way the dark ones, or so affect them that they cannot function properly?

Ashtar: "I am able to answer that in the affirmative, to this extent. The forces I command know that certain evolutionary courses have to be followed by your planet, and it is unfortunate that destruction is a part of this evolutionary plan. There is no way that this particular facet of human conduct can be eliminated except by evolution. Therefore, our concern is not so much with explosions, bombing or killings but with the fission of a living substance, such as hydrogen. This is not permitted, and we will halt any experimentation of this type. No living substance may be fissionized. The effect of atomic bombs on the dark forces is very profound. The particular form of blast caused by nuclear or atomic weapons, sunders the intelligence faculties from the vehicles (i.e. bodies) they are associated with and scatters these into space where they cannot be reassembled. Thus, the atomic bomb means for them annihilation, even as it will eventually mean a similar fate for humanity if employed to excess, at least as far as physical life is concerned. The effect upon the particular molecular structure of the dark ones is profound. Therefore you find that practically all communications from them deal with and dwell on this particular thing. It is in a sense a manifestation of the Power of the Almighty and it is curtailing their activities considerably. It is also a weapon against them, for which they have no answer. How strange it is after all their depredations that they should suddenly ask someone to be merciful to them."

The reader must decide for himself regarding the merit of this particular communication. Almost all philosophical and occult writing freely available on the subject indicates that matter on the astral planes disintegrates easily but comes together, i.e., reassembles, very quickly. Could it be that this bomb's effects interfere with this reassembling of their bodies as stated above? Only time and expanding knowledge can prove this, but this communication does have the supreme virtue of originality.

More important even than the suggested effects of these bombs is the fact that our whole nuclear scientific structure on this planet, while pursuing one obvious course, i. e., the destruction of the other fellow, may well be nothing more than a tool in the hands of the Almighty to work out a plan laid where plans do not misfire.

Not even that member of the Infallible Phalanx who crowed after the launching of Sputnik "No sign of God has been revealed in outer space" could deny that when he looks upon that fearful, searing fireball, surging with incredible blasting force, it epitomizes a power and a might beyond man's present station.

In its brilliant splendor, irresistibility and power, does it not suggest the Power of the Almighty? As with all else, time will tell.

CHAPTER SIX

ARE THERE ASSAULTS FROM THE INVISIBLE?

The concept of aggressive forces among the UFO in addition to benign UFO has been fairly conspicuously dealt with in this book, but without any evidence to support these communications from the invisible, we are just writing so many words.

It is clearly irrational to suppose that any beings living in space, or who have "conquered" space *must be* advanced, kindly, and intent on our rescue. We know only of the relationship of science to morality as it applies on our own planet, and this in itself should be a guide to what we might expect from some of our eerie visitants.

For the past seventeen years, a large proportion of our worldly wealth has been expended on mastering the "art" of casting down from the skies the means of destroying whole nations at one blow. This activity is still the major pre-occupation of the "leading" powers of this world, in this age of imminent space travel. We have behind all our efforts at stepping off into space, supremacy over that nation which happens to menace our ideology.

It is therefore irrational to conclude that any and all beings already in space are automatically wonderfully kind and filled with love for us. They may be even more deteriorated morally than we are, since deterioration of morality has gone hand in glove with scientific advancement, in our case.

In this world, we have good, bad and indifferent persons, nations and alliances, with the good and bad themselves, depending upon the viewpoint. Not only is it irrational to expect that other worlds, visible or invisible, will be any different, it is the acme of gullibility.

My own thoughts pondered many months upon this sombre reasoning, and upon the many instances of hostility or attempts to terrorize dealt with by Mr. Harold T. Wilkins in his UFO books. Mr. Wilkins, after sifting through piles of material covering many centuries has recorded many of these happenings, and also pointed the finger at a host of contemporary events connected with the UFO where aggression could have or may have been involved.

To Mr. Wilkins' work, we might add that of M.K. Jessup, the first astronomer to see the light, who has recorded many startling stories of teleportation, abductions and other out-of-this-world phenomena.

World events today are moving at an ever-quickening pace. The tempo of world history as it unfolds has accelerated greatly over pre-World War II days, and compared with Victorian times we are in a mad gallop. Crisis seems to be in the air, as Asiatic multitudes stir and lay hands on the tools of the 20th century.

It seems to me that when reason becomes as bankrupt and wayward as our science, man might *administer to himself* the final lesson in the necessity of practicing the Golden Rule.

In any such happening, we might expect participation on the part of those to whom the Golden Rule is anathema. In this frame of mind I addressed the following question to Ashtar.

Question: What are the present intentions of the dark ones concerning assaults on our surface?

Ashtar: "During the next few weeks and months, the astral forces will increase their activities along the western coasts of America. Many of their craft will be sighted. Should aircraft of yours be sent against them, they will be destroyed by the agency of a heat ray. Fires of mysterious origin will occur in the Americas and Western Europe. Explosions and unforecasted weather phenomena are part of their system of attack. Communications will be interrupted and airplanes forced down. Failures in electronic equipment are easily induced by them."

This communication, received July 17, 1956, can be verified in its prophetic aspects by a search of the newspaper files from that date. Suffice it to say that all the events mentioned occurred, although there may well be official explanations that are quite satisfactory. Such communications can only be weighed on their merits, and in order to do that, it is pretty well necessary to admit to the presence of hostile forces to check the events forecast. There is little harm in doing this on a tentative basis.

With this reference to a heat ray, it seemed to me that Buck Rogers was coming home to roost. To preclude the possibility that such things could be used is the worst kind of dogma. Super performing craft can be expected to have super performing weapons, even as our own U.S.A.F. jet interceptors now carry the incredible "Gatling" type cannon in the nose. Once more I asked.

Question: Is this heat ray the one which has been reported in some attacks on earth aircraft?

Ashtar: "Yes. It is a heat ray which has been used to interfere with ignition systems and to start extensive fires in various parts of the world."

The reader must be puzzled as I was as to why the attacks are carried out, and also how, if these weird entities are of a different order of matter to our own, they could interfere with our matter? Also, how could they escape detection? Once more, I asked:

Question: Many investigators are anxious to know *why* the dark forces cause these crashes of planes and collisions involving loss of life. It is not readily apparent to us, nor exactly how it is done.

There was a considerable pause before the answer, as though this being were wondering whether or not he would tax our credulity too far. The answer when it came, was jarring, almost shocking, but once again it fitted the facts.

Ashtar: "First let me say that the dark ones are highly desirous of causing destruction to airplanes and do so frequently. They would, if it were possible, bring down many more of them if this action would not result in their being detected by some of your people. The reason for this airplane crashing is really very simple for one who has grasped the concept of astral and etheric realms beyond the physical. Frequently, the astrals desire some particular person for a special purpose. Perhaps a technician, an engineer, or one of special skill or talent. They wish to abduct him, in other words. After an airplane crash, when the person concerned is released from his physical body after his 'death' his astral form is seized and taken to the nether regions. The crashes are brought about by several agencies. First, instrumentation failure, resulting in collision of one kind or another. Secondly, production of fire, usually in the vicinity of fuel tanks. Thirdly, complete suspension of the entire electrical system of the airplane. Fourthly, the use of the force field of their own craft to induce structural failure."

This statement is absolutely loaded with meaning. However, I had already found this particular being to be truthful insofar as I had the ability to find out what the truth was. The manner in which I might garble, or misinterpret what was offered must also be allowed for, in deciding if this unseen individual offers truth or falsehood.

Let us now get down to examining this series of communications in the cold light of evidence and common sense.

First, it is obvious that the best method of concealment for these activities lies in carrying them out away from the gaze of human beings. That is perhaps why so many mysterious airplane losses take place out of sight of land over water, high in the mountains, or anywhere human beings are not able to observe what transpires. Aside from the protection afforded them against the truth that they are present and doing these things, acts performed in clandestine fashion are covered up by another subtle thing. That is that anyone advancing the theory that they *are* done is immediately classified as being 'off the beam.' As the investigator in the old Universal Pictures film "Dracula" said in one unnoticed line: "The strength of the vampire lies in the fact that nobody believes he exists."

These covert, cowardly attacks are altogether typical of the forces purported to be inflicting them. They are the hallmark or signature of the forces carrying them out, and are the "fruits" by which they should be known.

Let us take first of all this "heat ray" allegation. Is there evidence that such a weapon is used on our planes in the air? I submit that the evidence is overpoweringly in the affirmative.

The well-known Walesville, N.Y., case is very definitely an instance of this. A Starfire jet fighter, chasing a UFO near Utica, N.Y. was *subjected* to a ferocious heat which almost drove the pilot and observer insane and forced them to bail out of their fighter. The jet crashed in the village of Walesville, with some loss of life. These airmen are fine young men, well-trained and equipped with the best that the United States can provide. To make such men bail out of a plane over a populated area, the heat had to be fantastic.

There was no question in this case of the UFO not being visible, for they were chasing it. It is also an instance of someone living to tell about what happened. Could there have been other similar instances in which there were no survivors?

What would have been the fate of these two men, for example, if the heat had been more intense and more applied? Probably instantaneous death followed by a fiery crash. It could be that there have been *many* more such attacks which have taken place away from human gaze, or in areas where the hostile UFO *thought* they were away from human gaze.

In the Walesville case, the UFO hovered above the scene of the crash, as large as life, *glowing brilliantly*. I ask the reader to bear these last three paragraphs in mind as I relate this next astounding UFO report.

This report reached me through the agency of Madame Manon Darlaine, a free-thinking and clear thinking lady from France who now lives in Los Angeles. Mme. Darlaine is one of a veritable army of sincere private persons who are probing the UFO mystery and who render priceless services to persons such as myself. This report was given to Madame Darlaine by Mr. Pierre Perry, of Beverly Hills, Calif., who has been a licensed pilot since 1911.

"As President of the Copper Mountain Mining Corporation of Arizona, I was on my way to inspect a mine in 1943 at a mountain location situated north of Prescott, Arizona. The mine is almost inaccessible. Close by runs the River Fria, and on a hot summer day around 5 p.m. I was leaving camp for the mine with two other men. These were another prospector and a Mexican named Isador Montoya from Marinette. We were on horseback, fording the river.

"All at once, Montoya, who was in the lead, yelled, 'El Diablos... El Diablos.....' (The Devils)

"We looked up and overhead a most terrific drama was unfolding, which lasted only a few minutes. A military plane was in sight, so were two large UFO's that looked like balloons (Montgolfiers) without baskets. They were luminous, and bright as the sun. The UFO's stood still, as though waiting for the plane to approach, then pounced towards it, at the same time projecting a violent luminous ray that could be compared with the large beam of a lighthouse.

"The air vibrated with a terrific explosion, as the plane was struck and came down. We saw two pilots bailing out, but as their parachutes opened, another fiery beam was projected from the UFO's. The chutes took fire, and the two helpless men fell to the ground to be crushed to death. The two bodies were later found.

"Meanwhile, the frightened Montoya was praying and crossing himself and repeating 'El Diablos, senor, I have seen them many times, senor.'

"Then from the horizon, coming from the north at an unimaginable speed, we saw another UFO. It joined the two above us, and together they shot like lightning to the south.

"We turned back to notify the authorities, and tell them what we had witnessed. Somehow they had already been alerted, and a truck and a jeep were on their way. We met them, and guided them to where we saw the wrecked plane had fallen. Parts were scattered all over the mountainside. It was a deliberate assault on one of our planes. I know it. At that time there was no talk of spaceships like there is today. Later I saw a sketch made by an airman, and recognized the same UFO's that I had seen that day in Arizona. I am a U.S. licensed pilot and a member of the Aviation Club of Savoie Aix-Les-Bains, France."

Mr. Perry's affidavit is to be found in the appendix. Through the courtesy of Mme. Darlaine I have met Mr. Perry and heard his story at first hand.

The sworn statement of Mr. Perry indicates that a ray weapon of some kind was used to shatter this particular airplane. This vindicates what Ashtar has offered us. The reaction of the Mexican miner, Montoya, is also of some interest. Not only had he seen the "same thing many times" but he characterized them as "El Diablos" (The Devils).

Strange craft which assault airplanes with heat rays, and slaughter defenceless pilots as they descend in parachutes, and do these things when and where they believe themselves to be unobserved are most definitely not of an ethical nature as we understand ethics here on earth. This action epitomizes unethical, satanic conduct, and these particular UFO entities must be known by their fruits.

I would like to add, because of its possible bearing on future "accidents" which may occur in this area that the official air navigation charts of the district warn of a "local magnetic disturbance" some twenty-five miles north of Prescott. The predilection of certain UFO to manifest around and near assorted magnetic vortices is well-known to investigators of UFO phenomena.

The Walesville case, and this particularly important sworn statement of a happening should be sufficient to prove that Ashtar was not fooling when he spoke of a heat ray.

Let us now examine the next method or attack alleged to be used, that of "production of fire, usually in the vicinity of the fuel tanks."

It is likely that this method would be an extension of the use of the heat ray, mentioned above, with the question of intensity to determine how much heat and therefore the extent of the fire.

In 1956, a Venezuelan airliner (Lockheed Constellation) got into moderate difficulties shortly after leaving the Atlantic coast of the U.S.A. The airliner, accompanied by a Coast Guard plane, was discharging gasoline prior to making an emergency landing. This is standard procedure and has been done many times in similar circumstances.

In this case, all communication with the airliner was suddenly cut off, and the gasoline caught fire, making a torch out of the airliner. The Constellation plunged into the sea with heavy loss of life.

The third system of attack, suspension of the entire electrical system of the airplane, is a little more difficult to prove, but there is evidence that these things can be done and have been done.

When one casts one's mind back to the days of World War II to recall that pilots limping back to

England after raids, in shattered bombers, were frequently able to make distress calls and get themselves picked up, one is led to wonder why it is that these disappearing airplanes under conditions of peace are unable to emit even a squeak of warning. Suspension of their electrical systems by means unknown is one explanation.

Harold T. Wilkins relates the following incident indicating that this interference can be carried on, and has been carried on before the world ever heard the term flying saucers. The excerpt is from Mr. Wilkins' book "Flying Saucers Uncensored."*

"In the war year 1944, an American pilot, flying over the Burma Road, said his plane was held motionless and the engine propellers stopped, while far aloft a mysterious disc appeared to be putting a sort of immobilizing ray on his plane. After this seeming 'inspection' his power came on again, his propellers resumed turning and the mysterious object disappeared into the far blue."

It is fatuous to suppose that all such incidents, and many more not exposed to the public gaze, are not in the files of Air Force Intelligence. This incident is quoted to prove that the information offered by Ashtar is not false.

On the question of the fourth method of attack employed, that of magnetic disintegration through the use of the UFO's force field, the now famous Mantell incident might find its explanation in this way. Reports of what actually happened to Mantell's plane as he pursued a giant UFO over Godman Field

*Citadel Press, New York.

Kentucky, have been distorted, and the truth lies buried in Air Force files. There seems to be little doubt, however, that the plane was "torn apart" by a force unknown exerted by the UFO. The Mustang fighter which Mantell was flying is one of the strongest airplanes ever built, all but impossible to tear apart in aerobatic manoeuvre.

Reports reaching me through devious but reliable channels enable me here and now to challenge the Air Force to issue an official statement, to which the Secretary of the Air Force will *swear under oath,* that no other incidents of a similar nature have occurred, and that no evidence exists of magnetic attacks on our planes.

Instrumentation failure as a system of attack could result in collision. There is very good evidence available that such tampering does go on, and it is to be found in the case of the six Navy planes which were lost off the Florida coast in 1946 without a trace ever being found of either the planes or the crews.

The last communications received from the pilots on the flight indicated that they were confused as to direction and that their *instruments were acting very strangely.* Shortly afterwards, communications were cut off and nothing more was heard of the five dive bombers. A huge Martin Mariner flying boat sent out to look for them also disappeared without trace.

There are literally dozens of examples of radio communication being cut off prior to crashes, and sometimes of its being cut off then restored just as mysteriously.

The recent Grand Canyon disaster (Arizona again) in which a DC6 and a Constellation collided taking more than a hundred people to their deaths has suspicious elements.

From the facts available to the public, two stand out like pinnacles. First, the sudden interruption of communications. One of the planes suddenly broke off contact in the middle of a sentence, while talking to the ground.

Secondly, the only witness to the disaster who was some distance away, stated that the planes "flew on stuck together." It seems inconceivable that under these circumstances the upper plane could not have sent out some kind of distress call.

An additional factor worthy of consideration in this case is the fact that the airplanes carried a large number of highly trained and highly placed people in the electronics industry, returning from a West Coast conference.

In an early discourse, Ashtar pointed out that these methods of attack which involved instrumentation failure left no trace, and therefore, our experts are often forced to ascribe these crashes to known and recognized official causes. Sometimes even the investigators are at a loss to describe some of the peculiar crashes that take place.

No one should assume forthwith that the Grand Canyon disaster was "caused by flying saucers" for such a viewpoint is not supported well enough at the present time. There are suspicious elements pointing to interference, but a recent "Life" magazine gives the full, "official" explanation of this disaster, and most people will be content with this.

We should remember, however, that unless observed, or unless there are survivors, these instances of UFO assault that do occur are not likely to leave any trace of their origin.

Regarding abduction of persons by the UFO, or their (the UFO) wanting certain personalities for their own purposes, there is evidence that bears out these somewhat radical concepts.

That these unethical UFO beings "desire some particular person for a special purpose" does not seem to be in any way unacceptable, especially when one considers that craft are appearing on our surface manned by very rudimentary types of beings. Perhaps skilled personnel are in demand. We don't *know* but we have the lesson of our own civilization which is that with the advance of science a voracious appetite is developed for skilled persons, one which cannot be filled fast enough.

In the United States today, we have men like Walter Dornberger, Werner Von Braun and Willi Ley, the former kings of German rocketry. These men all enjoy U.S. citizenship, presumably without answering the question required of the tattered refugee, "have you ever sought to overthrow the United States Government by force?" These men, who little more than a decade ago hurled rockets against the free world, now serve that same world, and we are glad to have them. So badly do we need their skills. Let us then beware of lightly dismissing the idea that UFO beings do not need anyone from this planet.

There is evidence supporting this view. A man who must remain anonymous but who knew very well one Karl Hunrath, a dark and mysterious figure in the early days of saucery, told me that Hunrath confided to him that "they need pilots." More than this, Hunrath would not reveal, but as Hunrath is now generally considered to have been some form of emissary for the dark forces, the conclusions are obvious.

It should also be remembered that Hunrath has since disappeared under very mysterious circumstances, with his trail leading directly to the UFO. The imprint of a triangle "pawnbroker" landing gear was found a few steps from his abandoned plane. He has not been seen since, nor has his companion on that last flight, one Wilbur Wilkinson, electronics expert.

If "they need pilots," might they not also need other scientific skills? We shall presently see how this all fits together. Certainly they could do worse than get pilots from this earth, where men have learned to fly pretty well.

The release of the person from the physical body, according to occult and superphysical students, does not deprive that person of earthly skills. They go with him. And since the physical body probably would not stand the vibration of the world where they are going, these beings doing the abducting do not seem to worry about murdering the personality they require. "Thou shalt not kill" is transgressed, and these are more of the fruits by which they shall be known.

Many persons will bridle at these concepts, but my suggestion is that they read the works of the late Max Heindel,* the great seer, who had the privilege of observing clairvoyantly what happens to the human body when it falls from a great height. It is illuminating and fits very neatly with what Ashtar has already presented to us.

For those who want more earthy evidence, there are incidents on record of abduction that need little comment. Whole airplanes are abducted, and this is not science-fiction balderdash. Here is a sworn statement from Mr. Eugene Metcalfe, of Paris, Ill. At my request Mr. Metcalfe made the affidavit which appears at the end of the book.

"On March 9th, 1955, at approximately 5:50 p.m., I witnessed the 'plane-napping' of a jet plane while standing in my backyard at Paris, Illinois. The plane was coming towards me from the southwest, and was traveling in a northeasterly direction. As I stood watching this plane, an odd looking craft came from behind it and just swallowed it. The UFO had an opening that was in my line of vision, and through the opening it took the plane. After this, the UFO hovered and pulsated and churned up and down. Then it seemed to whirl and lift upwards. While going through these gyrations, vapor came from porthole-like openings around the bottom part. The plane and the UFO were in perfect view, and stood out clearly against the sky. The object was bright

---
*Obtainable from the Rosicrucian Fellowship, Oceanside, Calif.

silver, and I heard no noise. The UFO was very big and bell-shaped."

Unpleasant? Of course. Shocking? Of course. But despite these aspects the sighting has great value for all people interested in getting at the bottom of the UFO mystery. It describes clearly what we consider to be on this earth a *highly unethical act*. It is a theft. Those who steal and who are unethical are *not* angelic.

When this amazing abduction is considered alongside the equally amazing Kimross incident, both happenings assume a new perspective. Mr. Metcalfe saw with his eyes what was probably seen by radar in the Kimross case. The most incredible thing about this whole business of airplane abduction lies in the fact that any person of common sense could view such happenings as being anything else but unscrupulous and unethical.

With regard to Mr. Metcalfe's sighting, we should note not only that this craft was *silver,* but that Paris, Illinois has recently been visited again by the UFO. A Pony League baseball game was stopped on July 18, 1957, just around sunset, by two objects following a jet fighter plane across the sky. Did they again have the intent to kidnap the plane as before? In any case, there were scores of witnesses to this latter event. When the jet pilot became *aware* of their presence, they took off in a hurry.

There are many apologists for these acts on the part of the UFO. These persons who apologize for and thereby endeavour to explain these happenings ought to pay due attention to the fact that they are in danger of becoming *tools* of these unethical beings, whoever they are or whatever they are.

The standard apologist's explanation runs something like this: "The saucers take pilots, and give them the choice. They say to the pilots, 'Do you want to go back, or do you want to remain with us'?" The pilots of course, many of them married with growing families and responsibilities, invariably choose to leave all their loved ones in the lurch and go sailing off in the UFO.

Aside from the obvious fact that in some instances the UFO may be claiming their own, it must also be observed that many of these abductions are deliberate interference by these UFO beings in the life of the person they abduct.

These sinister, mystifying cases go on without surcease. Word recently reached me of yet another abduction of pilots from a jet trainer out of Bolling Field, Washington, D.C. The trainer took off with two occupants, and three hours fuel. Given up for lost, the jet trainer suddenly appeared over the airfield six hours later, landed and skidded to a halt at the end of the runway. *There were no occupants.*

Obviously, the plane could not land itself, and must have been either set down from an invisible craft as described in this book, or else it was manned by spirit beings exercising control over matter in a manner similar to the well-known poltergeist phenomena, and having the required skill to pilot an airplane.

Its virtually impossible to explain this incident any other way than by the presence of invisible intelligences. There are doubtless other similar incidents which a puzzled Air Force will not divulge. The story of this particular happening was related by George Adamski, in a letter he read at a 1957 lecture in Van Nuys, California.

This periodic abduction of planes, crews and individuals seemed to me to call for another question to my source. On this occasion I was switched by what was purported to be a special transmission link to another being named Andolo, himself purported to be aboard the vast satellite Shan-Chea mentioned earlier. Describing himself as a "Member of the Council of Seven Lights," the impression of him I gained was that of a very old man, wise and cautious.

Question: We have been puzzled by the abductions of whole airplanes and crews periodically. There are some things we would like to know. First, do you ever for any reason take human beings off this earth, or abduct them so to speak? Secondly, why are these airplanes taken?

Andolo: "First, our laws forbid us to interfere in any way with the life plan of any entity. We cannot bring about a physical death wittingly under any circumstances. Everything must proceed in its Universal plan, which we are bound to carry out. However, the dark ones do not live by these laws, and prey upon mankind freely, seeking vehicles for themselves by any method they know. They do not hesitate to bring about the 'death' as you term it, of any entity if it serves their purpose. We have nothing to gain or learn by taking your machines. Such activity is meaningless to us. Not so the dark ones however. They have several reasons for the abductions of airplanes and crews. First, they may desire the actual physical material of which the machine is constructed, for purposes of their own which don't concern you at the moment. Secondly, they may desire the entities in the airplane for purposes of their own, regarding which I shall presently tell you nothing. Thirdly, they may desire both these two things in combination. The advancement in flight of earthmen is moving steadily towards the 3500 mph which is the limit of their present attainment. They are interested in seeing what you are doing, and how you are doing it. The increasing mastery of the science of flight, soon to make extra large forward steps is concerning them greatly. All this is a little beyond your present understanding, but eventually you will piece it all together and know the whole Truth."

Here at least a motive is offered for what has been happening. An interest in our mastery of flight as we move into the Air Age. This communication was received in September of 1956, and it is unquestionable that the science of flight has bounded forward tremendously since that time. Since the communication was received, "extra large forward steps" have indeed been made. Here is the proof.

Lieutenant General C.S. Irvine, Deputy U.S.A.F. Chief of Staff for Materiel, speaking in Fort Worth, Texas on July 9th, 1957, described the "sandwich" construction technique used on the B-58 Hustler bomber as "one of the most vital achievments of recent years." General Irvine also stated that this same technique may turn out to be even more important for the future development of missiles and advanced aircraft than it has for the B-58.

Another advance of great significance in mankind's march towards conquest of the air is the discovery by University of California researchers that projectiles travelling 760 mph get hotter at altitudes *above* 20 miles than at lower altitudes. Conventional theory had previously held that friction from the dense air at lower altitudes would produce greater heat. This discovery was announced in August 1957.

As to the desire of these unethical entities to utilize the skills of those they abduct, this is an answer that time and an extension of the understanding will provide. At least in the meantime it is better than *no answer at all.*

Perhaps with the idea that I should be given something more positive and concrete to bear out the somewhat jolting information they were feeding me, the following information was given to me on my birthday, September 17, 1956. Because of its nature, I immediately conveyed its content to Dr. Franklin Thomas of the New Age Publishing Company. In addition, my friend James O. Woods was present when the communication was received. It amounted to a virtual invitation to verify their information.

Ashtar: "I wish to alert you to the forthcoming renewed attacks on your planes which will be carried on by the dark ones. Keep a careful watch on these reports."

This communication permitted almost immediate correlation of the telepathically imparted information with events due to transpire in the immediate future. I resolved that I would watch the news with great care, and see what developed from the date of this advice. I was disturbed by what happened. The reader must decide for himself whether or not there is any value or virtue to what I have recorded The newspapers of the Los Angeles area are the only sources for these reports. They were sufficent. There may have been many additional mishaps in other parts of the world not considered newsworthy in Los Angeles and therefore not published. Here is the record:

Sept. 17: Lockheed U-2 high altitude research craft reportedly exploded and crashed near Kaiserlautern, Germany. The J-57 powered jet was in Europe conducting tests on clear air turbulence, convective clouds and jet streams at altitudes of 50,000-55,000 feet. The pilot was missing and feared dead.

Sept. 18: Madera, Calif. B-52 Eight-jet bomber bursts into flames in midair. Five men killed, pilot escaped, together with another man variously described as a maintenance expert, etc. Earlier in the year (1956) another plane of the same type exploded in flight near Tracy, Calif. with *malfunction of the complex electric system* blamed for the crash.

Sept. 22: Eindhoven, Holland. Jet fighter crashes into thickly populated area. Pilot and civilians killed.

Sept. 22: San Bernardino Cal. Jet fighter crashes on Mt. McKinley in San Bernardino mountains.

Pilot escapes, but crash starts worst fire in history of area.

Sept. 24: Mt. Yale, Col. Air Force C47 crashes into the north face of this mountain at 10.30 am. Twelve occupants killed instantly.

Sept. 25: Tallahassee, Fla. Two jets collide over the city with all occupants killed in both planes.

Sept. 26: El Toro, Calif. Crippled F9F Pantherjet makes crash landing. (No mention of what "crippled" the jet.)

Sept. 26: Three army planes suffer disabled engines over the Pacific within two hundred miles of each other and within three hours of each other. All planes suffer similar engine malfunction but all get back to San Francisco.

Sept. 27: Laredo, Tex. Four fliers killed as two jets collide over Laredo Air Force Base.

Sept. 28: Fallon Islands, Calif. Navy Skyraider crashes into sea while on a flight from San Diego.

Sept. 28: Austin, Texas. F84 Pantherjet crashes on base, pilot killed.

Sept. 29: Edwards Air Force Base, Calif. X-2, 1900 mph test plane crashes near base killing pilot. "We had radio contact with the X-2, but the contact suddenly stopped and we don't know what happened." Statement by Colonel Albert A. Arnhym, Public Information Officer.

Oct. 1: London, England. Britain's Avro Vulcan bomber, returning from a highly successful tour of Australia and New Zealand (a 26,000 mile flight) crashes and explodes on landing.

Oct. 2: Tokyo, Japan. USAF KB29 tanker plane crashes. Got out of control on landing.

Oct. 2: San Diego, Calif. Navy AD5N Skyraider crashes into San Diego bay, pilot missing.

Oct. 3: Nome, Alaska. Alaska Airline's plane crashes in snowstorm, five killed.

Oct. 3: Honolulu, Hawaii. Marine jet fighter missing on night training flight. Fury jet was last heard from approaching its base at Kanehoe Naval Air Station.

Oct. 3: Charleston, S.C. C-124 Globemaster crashes and burns on landing at Kelley Air Force Base. Coming in with one dead engine, plane lost power on second engine during final approach.

Oct. 4: Rapid City, S.D. An F-86-D jet based at Ellsworth Air Force Base crashes. Pilot bails out, after "feeling an explosion in the tail of the plane." Live rockets are strewn around crash site.

Oct. 4: El Cajon, Calif. Navy F3H Demon Fighter crashes on mountainside. Pilot bailed out.

Oct. 5: Everett, Wash. Two F89H interceptors missing, believed to have collided in flight. Both planes carried pilot and radar observer.

Oct. 6: Medecine Bow Peak, Wyoming. DC4 airliner crashes enroute from Denver to Salt Lake City. The Civil Aeronautics Board, in its formal report, said the probable cause was pilot deviation from the *planned* Denver-Salt Lake City route for some "unknown reason."

Only a fool would point to such a list of accidents and say it is concrete proof of interference by UFO or proof of the hostility of some of them. It is *not* concrete proof, but it *is* a highly unusual list of accidents when it is considered alongside Ashtar's message of September 17,1956.

Many of these strange happenings in the air may be part of a long term plan to terrorize, i.e. to produce fear in people. Referring to this matter, Ashtar made the following statement:

"Let me give you this earnest injunction. Cast out all fear, for fear is the lubricant, the food and the vehicle of the dark forces. Without fear you are truly impregnable, so strive for this attainment."

This would seem friendly and wise, although one ought to beware of suggestions from any invisible being, no matter how friendly he may be, that the entirely natural function of fear, which is to *alert* a person to danger be abolished. What is more probably intended in this communication is the idea that knowledge provides mastery of fear.

The establishment of supremacy over what one may term the evil side, according to his individual light, lies in elevating oneself more fully into the opposite, or good side. The appearance of the UFO in recent years, and their accelerated activities in contemporary times indicate that mankind is approaching a point where he must choose. He cannot go upwards and downwards at the same time. Aspiration towards the better things will automatically impel man's evolution away from vicious veldts of destruction where he now seems wont to lurk.

As for the seeming contradiction found in the possession of extremely advanced scientific devices by certain of these degenerate spirit beings, who appear to have aggressive designs on some aspects of our civilization, let another statement from Ashtar be our guide:

"Scientific advancement is not now, never was and never will be related to moral development."

Let those who adhere to the fiction that the advent of the UFO is purely scientific phenomenon, unrelated to spiritual, moral or ethical values ponder upon the self-evident truth of this statement.

CHAPTER SEVEN

DON'T CALL US, WE'LL CALL YOU

Beloved, believe not every spirit, but try the spirits whether they are of God: because many false prophets are gone out into the world.
Hereby know ye the spirit of God: Every spirit that confesseth that Jesus Christ is come in the flesh is of God.
And every spirit that confesseth not that Jesus Christ is come in the flesh is not of God: and this is that spirit of anti-Christ, whereof ye have heard that it should come; and even now already is it in the world.
— John 4: 1-3.

The problems of communication presented to ethical beings seeking to contact earthlings are very great. Although they appear to command means of communication that are beyond anything possessed or envisioned by us, vast care must be exercised in its use.

To begin with, there are probably very few people at the differing educational, social and intellectual levels of earth life who have any value to these etheric visitors as *conscious* helpers. From among these few, only a minute percentage have the appropriate knowledge of the UFO phenomenon, desire to serve and physical capacity to withstand the tremendous vibration which these beings seem to radiate.

My own early experiences were not always pleasant, for, as indicated earlier, a form of strain is placed upon the body until it becomes accustomed to the vibrations of the devices employed to carry on communication. Any person with a weak heart, or organic troubles of any kind is unlikely to be subjected to the slightest physical risk by ethical personalities seen or unseen. There is evidence that Ashtar and others like him know our physical condition with exactness, and it is probable that they read it from our auric emanations.

In recent years, since the first contacts with spacemen were made, a great deal of indiscriminate trance mediumship has been indulged in by untrained persons of varying scruples.

The result has been a plethora of communications from alleged "spacemen," many of which are of a specious nature, self-contradictory and at variance with the observed facts. The point has now been reached in the various "saucer clubs" where the sound of a voice, *any voice,* on a tape recorder, mumbling about "the confederation of planetary governments" or other questionable things is greeted with delighted cries of "It's the spacemen speaking." Voices of unseen personalities, no matter who they may *claim* to be, are no guarantee of truth. The thing that makes a man a liar is that he pretends to tell the truth and takes you in! A skillful liar often gets away with lying for years in our own society.

Another difficulty lies in the fact that many people claiming contact with "spacemen" want to be the new Messiah. Many believe that they are *the* one, specially selected, painstakingly chosen from amongst all humanity for magnificent work. In all too many cases the intellectual level of the persons making such claims, together with their total lack of discrimination, merely indicates that they are ideally suited to be fooled. They are set-ups for those unethical invisibles who are themselves set up to fool, to delude and to confuse.

The particular being referred to in this book, Ashtar, once expressed the view that ALL communications, no matter whence they are purported to emanate, be considered solely on their own merits. He has frequently stated that many come IN HIS NAME AND UNDER HIS NAME, and therefore that it is foolish to presume that because a message comes "from Ashtar" it is automatically the truth. It is his wish that all messages be scrutinized and given the full glare of critical judgment regardless of the purported source, and especially of the *name* employed to gain confidence and a ready ear.

An "Ashtar" message may not be from Ashtar at all, but from one of the scores of impersonators, masqueraders and deceivers who abound in the invisible worlds just as they do here. Why should the invisible worlds not contain these types in the same way our own world does? The dockets of our courts are filled for months to come with cases involving delusion, bilking, embezzlement, larceny and similar crimes, and it is illogical to think that in all worlds beyond our own all purveyors of such negative qualities have ceased to exist. These things constrain wariness, skepticism and caution. And no ethical being should resent having his communications scrutinized in this way.

As a demonstration of how unethical some of these invisible personalities can be, in their attempts to convey information to our world, I know of one quite harmless and God-fearing man who periodically and without his own consent, goes *under the control* of an invisible entity. That is, every faculty of this man is surrendered to the control of this invisible being. Acting without discrimination either towards this process or towards the information imparted by the "control," the man is fully convinced the "spacemen" are communicating through him.

It is reasonable to presume that we are placed in our own world, and for the most part, sealed off from these invisible worlds for a reason. The reason lies impenetrably as a function of the Divine Mind. We can easily reason that if this is the case, the veil between these worlds ought not to be broken without serious deliberation and never without preparation and training. Consider the parallel again to be found in our own earth life.

Radio operators have to be sure which station they are communicating with before they pass messages. They do not send messages to the wrong place, but select the station they wish to contact through *tuning and a process of identification.* They are trained to do this.

Identification must be exchanged to the satisfaction of both parties before intelligence is conveyed, and telepathy and other communications should not be regarded as being any different. In fact, greater care should be taken because one's *mind* is concerned in this form of communication, not material equipment.

Ethical beings wishing to convey information will usually ensure that the person they contact will have access to and will seek the protective measures necessary to prevent the unethical beings from interfering with and *corrupting* the channel they wish to use.

For these ethical beings to open these faculties in a human being, probably by etheric means, is therefore a very solemn undertaking. Naturally, for their own purposes, they do not want contacts on earth who are just "broad tuned" to everything thrust at them from the invisible. They want their own instrumentalities, who can be available to them as near to exclusively as possible. The opposition, on the other hand, wants as many channels as possible into which they may pour their cosmic drivel, in order that their own activities remain camouflaged by the confusion they generate. Additionally they want tools to perform earthly functions which they themselves cannot hope to perform.

Here is a warning on this matter from Andolo, mentioned earlier in the book:

Andolo: "I shall issue this warning. Those of you who are seeking to make contact with the higher forces, are exposing yourselves to dangers from which it may be beyond our power to deliver you once they have happened. Our suggestion to all those seeking contact by other than mechanical or artificial means is to first seek protection and knowledge from what are termed on your surface, 'occult sources'. If you yourself rend the veil between yourself and the invisible worlds, then you may find waiting on the other side, forces alien to all your ideals and higher concepts. These forces will be unwilling to release you from their toils once you are in them. I therefore enjoin greatest care and caution in attempting telepathic contact with us. If you will but read the earlier communications of this channel you will realize as nothing else will make you realize the type of creatures with whom you may be making contact. Some of them are highly evolved, and are able to masquerade as spacemen quite effectively. Others have nothing valid to impart, and use your contact with them for joking amusement or as a means of obtaining control of a physical vehicle. These things deserve to be weighed and evaluated most carefully by those, who, however sincere their beliefs, may in the very act of contacting us be exposing themselves to all that they do not wish to contact either in the visible or invisible realms. We have the power and means to make contact with any person we desire to be of *conscious* service to us, and when we make such contact, it is made so that the person concerned is neither harmed nor placed in danger."

To my mind, this communication contains an abundance of pure common sense, and I unhesitatingly commend its entire content to all would-be dabblers in these matters.

I know from personal experience that much of what Andolo has stated is the truth. Much of what he says is self-explanatory, and as we mentioned earlier, we cannot arbitrarily exclude from the possession of telepathic ability, such beings as the "Amoeba" or the "Peekers." Turn to the illustrations and look them over again. Compare them with the communication and realize that in this case, a friendly warning has been offered from the invisible.

When the faculties are first opened to these thought communications, one is continuously bothered by a succession of masqueraders who seek admission to the consciousness to have their "two bits worth." If one becomes physically tired, these visits tend to become assaults, and only vigorously applied knowledge repels the invaders. As Dr. Alexander Cannon says, "One does not get rid of evil by namby pambyism."

Communication may take many forms. In my case, the entire vocal mechanism was operated by the invisible communicator, without my losing consciousness or the power to interrogate the invisible being. This method seems to come more readily under the control of the receptor than the others.

The receptor may hear a voice inside his head, seemingly, and be able to vocalize the intelligence so received. Or the receptor may hear voices seemingly from outside his personal world. At the first sign of these manifestations, the person concerned, if untrained, should seek advice immediately, for such things lead on to often disastrous consequences. Not the least of these is the receptor being convinced by mischievous entities that he is Napoleon or the President of the United States.

In California today we have a group of young women "receptors" who have gone around impressing their friends with complacent statements about being used for "tests" and "wonderful experiments" with the "space people." Tests and experiments of this type usually reach the point where even the feelings of these women are under unseen control, and "space" craft land, disgorging giant entities who bang on midnight doors.

One must resolve to keep all such unsavory invisibles out of one's life and personal world, even if the "penalty" be no communication at all. The prize is the security of one's own temple, beside which the reception of trash from invisibles is not to be compared.

Religion has often been considered to be inextricably involved with the flying saucers, although there are some who peddle the line that the whole manifestation is a scientific event with no spiritual overtones at all. Religion in the sense of sectarian, confused beliefs probably is not involved, but spirits and spirit forces as well as spiritual values, are most definitely concerned, in my opinion.

When I do business in this world with another man, it worries me not one jot what faith he follows, because we are compelled to behave in accordance with the laws of the land in which we live. We obey these laws and our business is transacted within them, and we get along very well together whether we be Catholics, Jews, Mormons or of any other religious faith.

When invisible beings are concerned however, who do not hail from our own plane of life and who live under laws that may bear no relationship to ours, their spiritual affiliations are a matter of prime interest to myself.

Reared in the Christian faith, I must insist that any invisible being serves the Christ Spirit before I will transact any business with him. In the stumbling, backsliding, groping fashion that is the hallmark of the Christian, I try to live according to the teachings of Jesus Christ.

Beings who share my spiritual allegiances will not evade suitable challenges, and satisfactory answers to these challenges were a necessary prelude to all my communications.

Even assuming these conditions are met, the receptor is still not freed of his obligations *to himself* to view all communications critically. After all, there are many Christians on earth who are nincompoops, even if well-meaning, so we are not freed from the necessity of using discrimination

I can attest to the fact that there are invisibles who are of the anti-Christ, and there may be many more who have spiritual allegiances of a kind we would not even begin to comprehend. After all, think of all the millions of followers of Buddha who have gone on into the invisible realms, personalities who are not inimical to Christians but who adhere to a highly intellectual approach to the Truth. An approach which shares the same truth root as Christianity.

Let a would-be receptor challenge his invisible communicators in the name of whatever allegiance he himself holds.

Let Christians beware of the fact that it is now almost two thousand years since Jesus trod the globe and that in that time many of the highly evolved and princely personalities of these dark realms may have learned by subtle and various devices to circumvent these challenges. The need for caution is ever present.

There may be many readers who will surge from their chairs at this point declaiming that religion is being forced down their throats and that all they want is the "facts about the saucers." Well, nobody is asked or expected to believe all that is in this work, but I felt that insofar as I have found the forces of the anti-Christ to be present, and that this fits in with Biblical warnings as mentioned at the head of this chapter, this belonged in the mosaic.

There is absolutely nothing to prevent any person with a highly developed critical faculty, who is not easily spoofed, from listening to any "spacemen" he chooses, or earthbound spirits, or other entities who communicate by any means whatsoever.

If the question of contact with these UFO is left entirely up to UFO beings, fewer people will lose themselves in the labyrinths of psychism, into which they are so precipitately plunging today in such large numbers. For the undiscriminating, the gullible, and the uncomprehending, attempts at contacting the UFO are better left alone.

Telepathy, for example, should be considered in the same terms as its humanly manufactured cousin, radiotelephony. Imagine the mind to be a receiving set, for the direct reception of thought waves, which are the counterpart of radio waves in the parallel I am drawing.

Not knowing how to tune your receiver properly, you nevertheless turn it on. Unable to tune it, and getting stations indiscriminately, you find also that you are unable to turn it off. Occultists are unanimous in their cautioning against such untutored meddling, so in radio terms, if you do not understand the operation of your receiver, do not turn it on, or force it to come on. In this way, you will not receive a lot of cosmic static, signifying nothing.

By remaining sealed in the physical world where he belongs, until contacted, the person concerned is rendering himself and these benign forces great service. They are not withdrawn from other more pressing matters, which could quite possibly be in our interests, to aid those who are both foolish and impatient. I know this problem of desire for contact very intimately, and have no trouble recalling the avidity and enthusiasm with which I sought it. I also have no trouble in recollecting the narrowness of my own escape.

What each person alert to the presence of the UFO should do, if he or she is thinking along the lines of contacting them, is to realize that the paths of service are manifold, and that service to the same cause may be rendered in many different ways, not exclusively by contact.

While much of what has already appeared in this book is well-known to initiates of certain esoteric brotherhoods, and to occult science students, generally, it is all probably new to the layman. The inquiring layman may wonder why it is that I have been used as a channel for the dissemination of this information. I wondered about this myself, and addressed a question to Ashtar on the subject.

Question: Why has the information you have given me been passed through me? That is, why have I been used as the channel for it?

Ashtar: "This information is well known in occult circles upon your planet, and has been known for many thousands of years by those you term initiates and adepts. This information, however, could not be revealed through their agency to the outside world. We therefore have selected individuals whose analytical minds and willingness to accept new concepts will make them channels for the truth. Your recognition of yourself as the channel for and not the originator of this material is both discreet and honorable."

Let it be seen from this communication that the knowledge advanced step by step is neither new nor revolutionary in those circles, where under oath, all these matters are studied and mastered by those who are *fitted* for such things. In my opinion, a great deal more is already known of these UFO by certain human beings on earth than could possibly be passed through a person of such limited abilities and knowledge as myself.

One must be very careful when invisible beings begin praising their receptors with such things as "you are discreet and honorable." While this may be true, it may also be the thin end of the wedge for piling on more and more flattery which is designed to make you lower your guard.

If you take the simple precaution of making a list of your personal qualities, both good and bad, being as dispassionate as possible in its compilation, you will find that studying it equally dispassionately equips you to identify any invisible liars who arrive dispensing fulsome praise.

Because so much psychically derived information on the UFO is contradictory, orthodox investigators who probe its fringes tend to discard all information so derived. They often reach the understandable conclusion that all psychically endowed persons are "nuts," not because of the fact that they receive the communications, but because there is precious little agreement or unanimity in what comes through.

The conclusion that appears to be overshot by almost every investigator dealing in psychic revelations regarding the UFO is that different *sources* are bound to give different information The Egyptian Government, for example, holds a viewpoint regarding its nationalization of the Suez Canal which is completely at variance with that held by Britain and France. The objectives of the parties involved are different, and so are the *sets of ethical valuations* they employ to develop their respective viewpoints. And yet both are talking about the same thing.

Transfer this state of affairs to the invisible realms whence these UFO communications originate, and you find that the same conditions apply. One faction is opposed to the other. Both are talking about the same thing. Both have different objectives, purposes to serve and different ethics. Just as we do not say that a difference of viewpoint means that both the British and the Egyptians are non-existent, so should we also beware of saying that differing viewpoints held by invisibles automatically refute the validity or reality of psychic information. On the contrary, the very fact that there are differences of an often almost diametric nature indicates the presence of at least two factions, and probably many factions. All of these factions and entities have their own ideas, their own axes to grind, and in many instances it seems that efforts are made to invalidate what comes from each particular faction's opposition.

Typical of this activity is the recently circulated rumor that Ashtar is some form of giant mechanical brain. Psychically derived information has made this allegation that Ashtar is a sort of cosmic Univac, into which the questions of stumbling earthlings are fed. Then, presumably, some long-suffering spaceman attendant punches a button and out shoots the answer.

I myself was suspicious of these "brain" communications not because of any great threat to the integrity of the communications I had been given, which are, after all, nothing more than just *information*. My suspicions were founded on the impression I had gained of this particular personality as being a somewhat stern military type of being, an impression that was borne out by his purported appearance in my presence to a clairvoyant lady of integrity in Los Angeles who knew nothing of my activities in this field.

Voice tones and pace of speech definitely reach the receptor in the method of communication used in my case, and while it is probably true that the personality communicating *used* a machine to transfer his thought, that *he, himself* was a machine seemed to be ridiculous. Here again I asked.

Question: I would like your comments on the rumor being circulated that Ashtar is a giant mechanical brain.

Ashtar: "I have in the past mentioned this to my other friends. This idea is sown maliciously to bring about confusion and disbelief in what is coming from this source. The dark ones, who are our foes, would very much like to see the faith built up in these contacts and messages betrayed and destroyed. And yet the merit of their content forbids any such thing taking place. Hence the idea of a 'giant mechanical brain' has been circulated to let your people think that they are contacting some vast mechanism instead of a *being*. I give you my personal assurance that I am as *real* and as *individual* as yourself. The fact that I am upon a plane of existence where higher vibrations prevail alters my reality not one whit. I do not see the reason for any further discussion of this point."

The reader must decide for himself as to the merit of this particular communication. Little needs to be added other than the fact that the "mechanical brain" concept originated, I believe, with a group of well-meaning but misguided persons which has journeyed around the Americas at the behest of mischievous spirits, dispensing trash about the end of the world purported to come from such exalted beings as the Archangels Michael and Gabriel. The plight of this group would seem to call for a liberal application of discrimination.

There is little doubt that in the minds of many people communication with the "space people" would seem to be closely related to spiritualism. This fact has caused many people to veer away from the subject, because of congenital religious taboos. The "etheric" concept of spiritual beings has made this confusion greater, and I felt that some sort of clarifying statement about the relationships between the various orders of beings would be in order.

Providing that one bears in mind that there are probably many vibratory levels or densities where these various beings dwell, the following communication from Andolo has some merit. Certainly no one is bound to accept it in toto, and it must be remembered that this being is endeavouring to convey something that is no doubt very profound to a mind (mine) that is, in most respects not sufficiently knowledgeable in these matters to readily grasp the significance of what the invisible being is trying to convey.

Question: I would like you to give your description of the differences between space people and spirit people. I would appreciate as full an answer as my understanding permits.

Andolo: "The differences are very great, although to the cursory glance it may seem that space people *are* spirits. However, it all comes down to a matter of the condition in which we dwell. We are etheric beings, in your expression. By this, I mean that we live upon a higher plane of existence. We are not discarnate in the sense of having *no* bodies. We have etheric bodies which are counterparts of your bodies but which are made of a more tenuous substance, and which are not subject in the same way to gravitational effects. The etheric state in which we dwell is one of many on an ascending evolutionary scale to which we all belong. Above us, for example, are beings more highly evolved than us by as great a gap as there is between ourselves and you. This is not meant in any derogatory sense towards you, but merely as a factual statement about the scales on which we dwell. Upon our plane of life we have much the same type of existence as you do, although it is free of the corruptions, crimes and undesirable elements which are to be worked out of a being's karmic life before he may pass into the etheric state. Your earth is a testing ground, one of many hundreds of thousands of testing grounds in the Universe, where beings evolve upwards on the scale of life, working constantly towards junction with the Great One as the ultimate attainment of all existence. We upon the etheric planes pass on to higher planes just as you do from earth, when qualified. This transformation on your plane is termed death. To us, it is merely a transformer interposed between us and the next level of life to which we will ascend. We will stress once more that the greatest factor in the way of a proper grasp of the true story of life on your planet is the conception of death as the end of all existence. Nothing could be further from the truth. Upon your plane, you must serve out an evolutionary period before you can ascend to higher worlds. The fleshly bodies which you have are part of a plan to aid your working through this scale of existence. At your level you must endure savage crimes, wars, strife and violence, and the Great One, in his wisdom, has seen fit to use the fleshly body as the most convenient method of taking care of this almost elemental stage of existence. Now then, there are more people waiting to serve out their karmic penalties than there are physical casings, or bodies to go around. Therefore, there is a dwelling place around your planet for these bodiless entities from your surface. These are spirits, or spirit people. They are, if you like, in suspension, or their evolution is interrupted, and they are anxious to return to the body either to carry out something they left undone, or else, after realizing the truth or partial truth of creation, to become incarnate again and work towards proper passage through earthly life prior to ascending to higher realms. We are in bodies of a tenuous but none the less real substance, vibrating at a rate much greater than that which prevails on your planet. Spirit people are dwellers in astral form who cannot go beyond that astral form without serving out their karmic life in a casing of flesh as you know it. Communication is possible, both with us and the spirit people. But in communicating with spirits you may find that they know but little more than earth dwellers, and in many cases, not as much. They may have nothing valid to impart. They may play tricks and jokes upon people contacting them. Therefore we can only suggest that all such communication be done with extreme care and *with a reason*. Without reason behind it, it is little more than folly. I leave you in love and goodwill. I am Andolo."

Some of my many friends in the spiritualist movement will perhaps not find some of these comments to their liking, but they are not reproduced here to please people, only to aid in getting at the truth. In anything connected with the superphysical, anybody can be wrong.

The idea of reincarnation is blandly presented in this communication as though any other concept is folly, which would bring this particular entity's attitude into line with the experiments of Dr. Cannon. Dr. Cannon, conducting over one thousand scientific experiments *to disprove reincarnation* proved conclusively instead that it is a fact, despite the opposite purpose of the experiments.

The reasons for our lives here on earth would appear to be much more complex, logical and intellectually acceptable than the combined hocus-pocus of doctor and priest would have us think, if this communication is valid.

Let the predominant lesson of this chapter be that all who seek to contact human beings by extrasensory means may not be spacemen at all, at least as they have been depicted as tall, handsome, with long golden hair and no dental caries. Along with beings who are supremely wise and beneficent, there are those who are malevolent and malicious, and those who may not live by any standard of values we could comprehend.

Telepathy, a form of psychism, can easily lead into thorny pathways if not thoroughly controlled from the beginning. It is my belief that psychism is a faculty that must be closed if one's further personal development is not to suffer. The constant seeking after "messages" from invisibles is a clear road to slavery. When something has been offered, as these communications have, to one completely uninitiated in occult and mystical lore, one is extremely lucky to be able to cut off the faculty as I have now done.

Realizing that I was venturing into realms where I was definitely a stranger, and in the presence of an order of beings I could not control, I decided to devote myself in the future to learning more about these things before opening any more doors.

One of America's most distinguished philosophers, Mr. Manly Palmer Hall, offers the following wise counsel on matters of this nature:

"It is most important that any student of metaphysics who expects to make any true progress in the development of his inner self should refrain from all forms of psychism, and avoid psychic phenomena."*

*Self-Unfoldment by Disciplines of Realization Manly P. Hall p.23.

CHAPTER EIGHT

ADVENTURES IN THE DESERT

Before the original booklet "Spacemen-Friends and Foes" was published, I had arranged to make a visit to the Mojave Desert to try and get some photographs of spacecraft.

In those days, I had not even begun to fully comprehend the significance of the information I had been given, and fondly imagined that Ashtar and his men would come down in their craft, become visible, I would photograph them, and that would be that. It was not that easy.

Although it took me many months to realize it, the etheric forces probably do not have the power to manifest at our frequency level, without the aid witting or unwitting of some person here. At the time I set my mind on the photographic project, Ashtar indicated that it would be difficult, but the paucity of my own knowledge undoubtedly prevented him from explaining exactly why it would be difficult.

In the year following this first experience, much has been learned about photographing other world entities but it was impossible then, as I now see, for Ashtar to explain all these things to my lowly intelligence.

Our first trip to the desert on this project took place on August 18, 1956, and our purpose was to get photographs *on purpose*. We have done the same thing many times since, and have driven many thousands of miles to master this art, or at least its elements. On this trip, we were a pair of splendid ignoramuses, and are appalled today by our own conceit in thinking that we were in just one weekend to lay bare the whole UFO mystery.

My good friend Jim Woods, whose affidavit covering all these events appears in the appendix, has been with me through the whole period of my investigations, and his friendship and financial aid have united with his natural aptitudes to make him a valuable partner.

So it was on August 18, 1956 we set out for the Mojave, planning to drive down the strip of desert between Victorville and Yucca Valley, and camp there in an isolated spot. Ashtar suggested that I be on the alert for an indication of where to turn off the road, but I did not attach special importance to this.

One the way to Victorville, the engine overheated badly, due to its having been dosed with acid solvent the previous day. I suggested to Jim that we pull into Morrow Field and drain out the loosened material that seemed to be clogging the cooling system.

At Morrow Field, I got out of the car to take off the radiator cap. Using a handkerchief and great care, I turned the cap slowly in accordance with the best practice. However, the moment the cap turned out from under the retaining lip of the radiator, the pressure was so great that it simply flung my hand aside and a geyser of dirty, boiling, acid-charged water shot into the air. I was caught on the right side of the face by this jet, and suffered a bad scald on the forehead, the cheek, nose and eye area. Mercifully, no large amount of water entered the eye.

In pain and suffering a little from shock, I thought the best thing to do would be to abandon the trip, for to drive into country hotter than that we were already in would have been unwise. So we turned around and drove back to Los Angeles, after stopping to apply medicants to my face and further attend to the overheated engine.

Arriving in Los Angeles about 7:30 p.m. we had dinner at my apartment, a painful process for me up until 8:30 p.m. At this time, we had pre-arranged a contact with Ashtar, but since we had abandoned our journey and were actually in the process of eating supper, we did not do so.

In the ten minutes following 8:30p.m. all sensation of pain completely left my face, to the point where I did not know I had been burned. I also felt a great renewal of strength, a surge of energy and power. I was even able to bang my knuckles against the burnt portion of my face, without any sensation of pain. Medical doctors can disbelieve this if they wish, for the purpose of the story is not to put forward new medical theories but to relate what happened.

We both suddenly found ourselves imbued with the desire to go to the desert, in spite of the fact that we had already driven over a hundred miles that day.

We made a point of prayers before we left, and set off again in very high spirits.

At 11 p.m. we were in Victorville, and could not help but compare our bouyant, zestful state with the depressed condition in which we had limped back to Los Angeles just a few hours previously.

Driving out beyond Victorville some nine or ten miles down the Lucerne Valley road, we saw a red light swinging in the sky above us. We stopped the car, pulled over to the side of the road and turned off both lights and engine.

Through twelve power binoculars, the object appeared to be hovering and dancing up and down slightly. We studied it for a few minutes, and were about to move on after chalking up an 'indefinite' when our attention was drawn to a monstrous cluster of colored lights moving across in front of us on the same bearing as the swinging light.

Through the glasses, this cluster of lights revealed itself to be a large cylindrical shaped craft, with red, green and white lights on it. I estimated its altitude roughly at 7,000 feet, and it moved from dead ahead to dead astern and out of sight in about thirty seconds. It was noiseless.

We were stunned by what we had seen. I thought it was probably a carrier craft of some kind, probably from six to eight times the size of a DC6 airliner. However, the completely clear sky and absence of noise made estimates difficult, because there was no way of gauging true altitude. It certainly seemed big.

We were certain it was an out-of-this-worlder a few minutes later when a B-36 or similar heavy bomber lumbered over at a greater altitude and the beat of its engines was clearly audible. In the silence of the desert, there is seldom any doubt of whether an airborne device makes any noise. Sound carries for miles.

We talked about the sighting for some time as we drove down the road to Yucca Valley. After about an hour of driving, I thought I detected a change in the beat of the engine. A few minutes later I was sure of it as the car began to slow down and the throttle began to close under *unseen control*. I floored the gas pedal, but the car would not respond, so I knew that this was the indication which we had been told might come. Gradually the hydramatic dropped down, until the car was crawling. Then the engine shut off completely and a gentle force on my arms turned the car to the roadside.

The engine stoppage was accompanied by a shout from Jim, who was studying the roadside. He spied a small road going off at a sharp angle from the main road. The entrance to it was covered with a pile of sand. As late as two months ago we had a great deal of difficulty locating this road in broad daylight, after having been over it many times. That we should have come to halt besides it at 1:30 a.m. in pitch darkness is certainly more than remarkable.

The car now came under control again, and I backed up and launched it on this bumpy side road. We jolted and lurched away from the main road for perhaps half a mile when we saw a wash ahead of us with low hills around it. We inspected it and decided that this must be the intended place for it was completely screened from the main road. A contact with Ashtar was next and his voice seemed tinged with amusement as he said: "It was with some trepidation that we discovered you at the appointed contact time at your home in Los Angeles and in pain from a burn on your face. *An etheric force* was utilized to withdraw pain from the burn, and make it possible for you to come here."

Subsequent studies by myself into various aspects of superphysical science, including occult principles of health, would indicate that such advanced beings would likely possess the powers of which he spoke. The only thing I can add to this was that I was very grateful to have been relieved of the pain.

Ashtar went on to tell us that we should be watchful around the dawn hour, and gave us the name of the "spaceman" who would be controlling the craft. He also suggested a means of protecting ourselves which must remain untold for the present. He concluded by stating that a *force field had been placed around us* by his forces. This was to have great significance in short order.

He promised nothing, but it is of great interest that all of our really good pictures have been taken in the immediate post dawn period.

Quite weary by now, we began devouring mugs of coffee. We had just finished our first mug when our attention was drawn to four objects which slowly materialized on the eastern horizon. Coffee went in all directions as we bolted for the car and began scrabbling in our gear for cameras and binoculars.

Through the glasses we gained the definite impression of the regular dome-shaped devices known as flying saucers. In particular, one manifestation to the northwest was especially bright. As we fixed our attention on it, it began slowly moving towards us, glowing a brilliant white color. At this time, I was not in possession of the light identification data contained earlier in this book, and immediately thought of this manifestation as being a friendly craft, what else? A little earlier than scheduled perhaps, but still our boy. This was not the only mistake I made.

Disregarding the careful procedures which must be followed in telepathy to protect the telepathist from phony or undesirable contacts, I immediately contacted the intelligence controlling the saucer, and he identified himself as the very spaceman Ashtar had previously nominated. We did not even stop to think that the scheduled appearance was actually *hours* away.

Excited beyond any regard for caution, I continued to converse telepathically with this being, audibly so that Jim could hear me. The intelligence said: "I am going to dim out the force field of this craft, then bring it back up to full brilliance again. Watch."

As we watched, the force field dimmed out to near invisibility then came back to full brilliance. Our hair stood on end!

Jim quickly set up the movie camera, although it did not have a film entirely suited to night work as we now realize. I continued in contact with the saucer entity, giving him directions to come over to us, and come down low so that we could get a good close shot of him.

He followed the instructions and came closer and closer until he was what appeared to be around six thousand feet above us and just to the east. Here he hovered.

Jim now had the camera ready, or so he thought, and just as he informed me of this, the saucer entity announced: "I am coming down now, watch."

The brilliant, bluish white light which actually hurt the eyes viewed through binoculars, now began its descent. It dropped perhaps five hundred feet, then seemed to hit an *invisible wall*. The craft literally bounced off this wall and shot sideways like a ping-pong ball. As we watched the saucer intelligence gathered his craft again, gained altitude and dashed his craft at us again. Again it hit this invisible wall only to carom off sideways, exactly as before.

Eerie, noiseless, brilliant, the craft was carrying out these manoeuvres under my constant telepathic urges to come down, that we would not be frightened and other invitations that I now recognize as being potentially disastrous.

After several attempts, all of which produced the same results, the craft ceased this operation, broke telepathic contact, and withdrew to a much higher altitude.

Almost prostrate with excitement, we were still nonplussed by what had happened. No doubt our fatigue had dulled our minds as well. What we did not realize until later was that we had seen a hostile or malevolent "space" craft (blue or bluish white light steady) attempting to make an attack of some kind on us. It had struck the force field that Ashtar had assured us had been placed around us.

It was not until the next day that we were able to piece the whole thing together, when communication with Ashtar revealed to us the series of dangerous mistakes we had made and the risks we had run. The whole experience was pregnant with lessons for me, and I hope others will learn from it. Great care and caution ought to attend UFO fans seeking contacts, for the UFO may not always be what they seem. A seemingly beautiful silver ship may be no more than a cosmic sheepskin for the entities within it.

This experience demonstrates the differences between two of the factions operating in our atmosphere, and of the need for wisdom and caution in approaching this vast and in some ways fantastic subject. There are foes as well as friends for each individual, and in this case, one of the most important lessons was the *masquerading* of foe as friend. The unscheduled entity presented himself as the being we were expecting. By not using discrimination, or common caution, we had unwittingly invited some possibly malevolent being to *come down,* an invitation they apparently seek with some avidity. The masquerading aspects of this incident, I have since discovered, are entirely part of the game when one deals with unseen beings. One must beware of falling into a similar plight to that of the trusting sucker in the gambling joint who suddenly yells, when it is too late, "I've been taken."

During the remainder of this remarkable night, we were treated to a splendorous display of spacecraft and assorted UFO which has never been equalled in perhaps thirty subsequent trips to the desert. At one time we had in sight, simultaneously, no less than nine UFO's, which we could identify as controlled objects. We heard an aircraft engine only once during the entire night, and we were easily able to identify it as a heavy bombardment type plane.

We saw one saucer shaped craft hurtle across the sky above us faster than anything we had ever seen, made by man, and likely to be operating above the Mojave at 3:30 a.m. We saw what appeared to be crackling trails of "angel's hair" etched into the black sky by craft or beings about their business. Several times we saw two or three craft come together in the sky and separate again after a brief period.

We had sought to make motion pictures of the machine descending towards us. The camera had jammed before the leader strip was even through it, and we could not free it. At the photographic shop where I took the camera to be unloaded in total darkness, it was discovered that the film had been wrongly threaded.

At this point, there will be those who will jeer at the idea of a camera jamming, and laugh the whole thing off. Jim and myself have subsequently gained mastery over various cameras, including movie cameras. A subsequent chapter tells the story of our photo efforts, but in this, our first effort, we were lucky to have the protection of friendly beings whose patience must have been tried on this and other occasions by our stupidity.

Early in September, 1956, we decided that we would head for the desert again and camp out over Saturday night in the hope of seeing something worthwhile in the skies. On the way to Victorville, about 9:30 p.m. Jim suddenly became incredibly sleepy. In spite of the fact that the top of the convertible was down, and wind cool he simply rolled into the back seat and soon was snoring heartily.

He slept about an hour and woke just before we reached our old location in the wash, which we did not do without difficulty in finding the access road.

He seemed refreshed, and we made a pot of coffee. We drank two or three cups, and then to my amazement, Jim found himself unable to stay awake. He just simply had to sleep, even though our purpose in going out into the desert was to stay up and watch. So he rolled into a sleeping bag and began a few minutes later to give a splendid imitation of a grampus.

Alone, and a little forlorn, I wandered around with the binoculars, looking for objects. Aside from a moving light far to the east, there was nothing. The sky was as clear as crystal, like a huge velvet cloth set with gems of purest ray. To the north I spied some moving lights, which might have been anything and I was about to give up my vigil when a strange thing happened.

Almost as though guided by an unseen force, I turned and looked into the heavens to the south. Above me in the sky, only about five thousand feet up, was what looked like a long black cloud, cigar shaped. It had not been there a few moments before when I had looked in that direction. It seemed nebulous, and yet solid.

As I watched, it split into four or five disc-shaped clouds of equal size which drifted away from each other and formed a rough square. Before my very eyes, the disc-shaped clouds drew together again into the cigar form and disappeared in an instant, as though switched off. Throughout this happening, I had felt goose pimples on the back of my neck, and I had been filled with a very unwelcome feeling of dread. I let this be my guide as to what type of a manifestation it was.

I was some forty yards from Jim during this happening, and while I watched the disc-shaped and cigar-shaped clouds, I was shouting to him to awake.

Towards the end of the manifestation, as the disc-shaped clouds drew together to reform the cigar, I ran to Jim and pummeled the top of his sleeping bag. He would not, or could not wake up!

Little comment is possible at present of an acceptable nature regarding Jim's inability to awaken, other than to point out that many strange happenings have taken place to one person when other persons present have not awakened. This was, in my opinion, quite possibly some form of demonstration of power, and judging by my own feelings in this case, no friends of mine were involved. Whoever, whatever these things were, their appearance was of great help to me in the later reasoning I developed concerning the nature of the reality of certain craft.

These two early experiences of ours should serve as a warning to people who, however sincere and enthusiastic, make trips to remote locales in hope of contacting strange craft. In the age we are now entering, contact is likely to become much more common, and people should learn to recognize the entities with whom they may have converse, telepathic or otherwise.

We have seen in earlier chapters how nefarious activities are carried on out of sight of humans, and this should reinforce the lessons of my experience. Other world entities of any kind, no matter how good they may *seem*, should be encountered only on our terms in places of our choosing. For the most difficult thing to learn in dealing with the UFO is that things are not always what they seem.

CHAPTER NINE

FRANKENSTEIN IN BLUE.

"This is a land of minority rule, where the few push the many."
— Senator William Borah, of Idaho.

Every drama has its villain. When the curtain went up on the UFO drama at the beginning of the "haunted decade"* there was no villain. So vast was the novelty of "extraterrestrial" visitation that nobody thought much about villainy. Not even the books of Harold T. Wilkins, who saw further and probed deeper than most, conveyed to the saucer multitude the idea of malevolence on the part of some of these visitants.

In most writings, the pre-occupation was with the means and methods of UFO propulsion, life on other planets and other similar guesswork. The craft were generally considered to be neutral, disinterested or benign.

Drama invariably imposes its requirements on those who participate in it, and so it was that soon an appetite arose for a villain. This happened because the UFO beings refused to state to all and sundry the true purposes of their presence. The absence of this revelation led to the fabrication of a villain, in

*Meade Layne.

much the same way as Frankenstein fabricated his monster — and with comparable results.

The saucer writers and investigators built their villainous monster, but instead of the seamless black smock of Frankenstein's creation, this monster was garbed in raiment of purest Air Force Blue.

The Air Force, by which generic term I mean the air forces of the world, became in the eyes of all the UFO fans a close-mouthed, uncomprehending, bureaucratic juggernaut, rolling inconsiderately over the rights of man. Men on platforms, thirsting for glory, accused the Air Force of everything from stupidity to treason, by way of deceit, cavilling and conspiracy. The idea took root that these airmen and those who gave them orders were expressly engaged in frustrating the public by suppressing UFO information.

On the other hand, there are those ten watt mentalities in every country who think that the air force knows *all* about the subject, that its statements are gospel, and that after all its "their job."

The truth lies, as usual, in between the two extremes. The UFO partisans wanted the Air Force to make announcements of a suitable kind, that is, suitable to them. Believing immediately that all these UFO were interplanetary and benign, an irrational viewpoint as now shown, the UFO partisans wanted something of that order from the Air Force. They wanted it known that our deliverance was at hand, through the agency of shoals of otherworld astronauts, golden haired and grinning. Armed with scientific gifts of incredible value, all they awaited was clearance to land. They forgot the Greeks of the Bible who "came bearing gifts."

In the early days, the natural forces of conservatism were aligned against the idea of the UFO. Arrogant apostles of orthodox science joined hands with jealous militarists to squash this "cult" that seemed to be growing up, so quickly and dangerously. In other words, the natural conservatism which properly should accompany all new discoveries was poisoned by arrogance and jealousy. Once upon this slippery slope, the Air Force has never since been able to get off it, even though it is well aware today of the presence of the UFO and of the malevolence of some of them.

This unfortunate process has gone so far, that I hesitate to speak out against it. In doing so, I am well aware that I am venturing upon a dangerous and creaking limb. Nevertheless, I believe that incorrect reasoning, inadequate knowledge, and understandable mistakes have united to cast a stigma upon the Air Force that ought to be removed. I do not say that the Air Force has always acted in good faith, "Project Bluebook" is a living testimony of that, but I do not believe that the Air Force has done anything that cannot be explained by ignorance and mistakes, usually in reasoning, and by its duty to protect.

The point has been reached where UFO fans speak of the Air Force with bared teeth, often abusing them almost as though they were our enemies instead of our protectors. It is my hope that this wrong can be set right, and the villainy credited to the proper forces, which are not the air forces of *this world*.

To understand at least part of the air force view, it is necessary to understand the military mind. The training of the military man is concentrated upon improving his ability to kill, and on increasing the means available to accomplish this. All the varied ancillary matters that are now taught in 'enlightened' military colleges are simply adjuncts to the fundamental purpose of killing. Attack and defense alike reduce themselves to the art of directing maximum killing power against the enemy.

When they are called upon to think, and think hard and originally as is required with the UFO phenomenon, the results are often not good. They are as lost in the field of original thought as Einstein would have been commanding the Normandy invasion, for, as Will Rogers put it, "we are all ignorant, but in different subjects." I hold that this is so regarding the military man in the teeth of the trend in modern America to catapult generals and admirals into high political office or to the top of the business tree upon "retirement" from the services, usually in their forties or early fifties. And the influx of more military mentalities into civilian posts as a result of nuclear age changeovers is a prospect before which all must blanch.

However, there is nothing in this that should condemn the military man, as such, for we have not yet quite reached the point where we can dispense with him. We must expect though, that when he is confronted with a set of conditions which are completely incomprehensible by all his military yardsticks, he will flounder. He is not floundering alone, for the greatest thinkers of our generation are up to their armpits in the same cosmic mire. It is a fact that the military has inherited virtually unlimited authority in a spiritual and spiritistic matter.

The situation from the military man's viewpoint is a difficult one. Craft suddenly rush at his planes from he knows not where, manned by he knows not who (or what), with an objective that orthodox reasoning cannot be expected to penetrate. Under these conditions, he would be doing less than his sworn duty as his country's defender if he did not clamp on some sort of a lid until time, intelligence and perhaps an expanding understanding helped him to figure out what was happening.

This sudden, alarming advent of super-performing craft created a problem for which the military mind, a materialistic mechanism, has not been prepared. When these machines of often vaporous, intangible substance suddenly began with assorted rays and magnetic devices to sunder his airplanes and steal them without written permission, the military airman became confused. The real wonder to me is that the morale of the world's military flyers did not crumble completely. It was at this time that their flinty training stood them in good stead. We can be thankful that they are so well trained, and despite what we may regard as the immoral nature of the killing process, we must realize that mankind has not yet learned not to kill. Until then, we need highly trained airmen, well armed and determined.

As far as the gathering of intelligence is concerned, the two major bodies involved are the United States Air Force, and the Royal Air Force with its Dominion Air Forces throughout the British Commonwealth. These formidable organizations of men and machinery, speaking a common language and girdling the globe, have as their sole *"raison d'etre"* the protection of the people they serve.

The intelligence services of these two nations bring to bear millions of dollars worth of specialized equipment and scores of thousands of highly trained men, some of them the possessors of brilliant analytical minds. The purpose of the whole organization is to detect in advance any activity against the nations it serves; to estimate data on potential enemies, and to see that this data, through appropriate channels, reaches the highest sources in the governments.

This massive organization, war proved and highly efficient, monitors our security. Into it, a plethora of data concerning the UFO has been systematically fed. Are we to believe after all that has happened that they have *nothing?* To believe that this is so is the acme of naivete!

Any politican or holder of high appointive office who expects me to believe that this far-flung international network of intelligence apparatus has produced nothing on the UFO is insulting my intelligence, which is only that of an average man.

In my opinion, the air forces hold in secrecy data that confounds the orthodox scientists they consult.

In my opinion, they have concealed from the public certain sightings, encounters and happenings connected with the UFO that have the potential ability to produce national panic were they to be released. I would be glad to see a sworn statement from the Secretary of the United States Air Force that all information concerning the UFO has been released to the public.

I cannot see how any UFO investigator can get excited over so-called official denials of these things when the official or military officer making such denials is not doing so under oath. Unless the person concerned does so testify, my reaction is to automatically conclude that he is telling a lie, even if it be only a white lie.

What would be the nature of these things that the air forces have taken such pains to conceal? The now famous Orson Welles scare of a Martian invasion contains the key to the truth. Orson Welles' over-realistic play did produce panic, and there is reason to say that any announcement along those lines by the Air Force would also produce panic, but this in itself could hardly be anything permanently injurious to the nation. Welles panicked the populace because his play dealt with an *attack* by Martians. Herein lies the tale. Attack, hostility, malevolence.

It is my opinion that the Air Force is well aware that it is being attacked by these craft of unknown type and origin. I personally believe that these attacking craft are aggressive spirit forces which owe allegiance to the Satanic deity, and that their bailiwick is our earth's envelope. I also believe, and in the future hope to prove that they enjoy the alliance of physical or near-physical beings from our own moon.

It is my belief that the Air Force presently does not have a defense against these craft, and that for obvious reasons of security cannot "tell the people" in the manner so energetically and inadvisedly urged upon them by investigators, UFO fans and saucer clubs.

Because of what appears to be pig-headed adherence to the desire for physical proof and therefore, physical beings who can be overcome by physical means, the Air Force has an extremely difficult job to perform. The professional education in killing is rendered grotesquely inadequate as high-powered jets fly right through the UFO, and as whole airplanes and crews are snatched bodily out of the skies by certain of the UFO. The Kimross incident and Mr. Metcalfe's sworn testimony of the weird happening at Paris, Illinois, are proof that these things do happen.

I believe that the Air Force has made some very bad mistakes, but they presently have all my sympathy in what they are attempting to do. The average businessman, if confronted with a similar series of obstacles in his daily life, would not like it at all. The ordinary worker, if confronted constantly with things beyond his comprehension would eventually be unhinged by such impacts. Within the limits of orthodox scientific knowledge, the Air Force is doing the best job it can, and only those who persist in clinging to the fiction that all the UFO are angelic will continue to drub the men in blue.

Returning to the question of the official statements made by Air Force Generals and lesser lights who share the joys of military discipline, people must realize that all military officers act in accordance with orders. The Air Force is not a place where you refuse to do something you don't like. The military services in general represent the last vestiges of slavery in the Western world, wherein one does what one is told, or faces dismissal, or punishment by court martial.

This state of affairs applies not only to lower ranks of officers, but even to such august and distinguished military men as General MacArthur. Therefore, all statements by air force generals, colonels and majors are just simply what they have been told to say. They may bear no relationship to the truth, but if *ordered* to do so, an officer must make such a statement or share the sad fate of General MacArthur.

Air Force General Sanford made a newsreel statement a few years ago which was eagerly seized on by scientists and others intent on proving the UFO to be non-existent. Unquestionably, General Sanford is a fine public servant, but in his capacity as a general he was not bound to tell the truth, but only to read what he was told to read. It is necessary to use great discrimination in analyzing all such statements. General Sanford stated that "no pattern" of hostility had been detected against the United States.

Had any hostility at all been detected, or just no "pattern" of hostility? Eighteen UFO's in V for-mation dropping fireballs on the Pentagon every night at six for two weeks would be a pattern of hostility, especially if the fireballs were to set light to some of the files and reduce the personnel needs. One UFO abducting an airplane over Paris, Ill. would not be a pattern. I cannot therefore accept General Sanford's statement as being either entirely truthful or exhaustive. He is, however, not to blame, for he stated what he was told to state.

The military officers at least have their disciplinary pyramid to excuse them. I am inclined to feel less happy about the civilian side of the governments. In my view, appointive office holders have a moral obligation to speak the truth or say nothing. As our morality deteriorates it seems that the sworn statement is the only recourse the public has to ensure that it does get the truth.

It is traditional in politics however, to speak evasively, to avoid the point while dazzling the interrogator with verbal rapier play. And always, in every statement, the politician leaves himself a line of retreat, a back door to the alley of escape. This is the art of the political announcement.

It is clearly shown in the comment on the U. S. A. F.'s "Project Bluebook" by Secretary of the Air Force Donald A. Quarles. Mr. Quarles said: "On the basis of this study, we believe that no objects such as those popularly described as flying saucers have overflown the United States. I feel certain that even the unknown three percent could have been explained as conventional phenomena or illusions if more complete observational data had been available."

Let us bring to bear upon this statement the same critical judgement we should employ with regard to communications from invisible intelligences.

In the first place, the Secretary plainly states that these *beliefs* are "on the basis of this study." Have there been other studies, carried on in secret, containing evidence not released in the "Bluebook" report? If "Project Bluebook" and other publicly announced projects connected with the UFO represent all the investigative effort of the U. S. Government into these matters, why should the Secretary leave himself the traditional political line of retreat by saying "on the basis of this study"?

Secondly, since the Air Force now holds certain beliefs about the "objects popularly known as flying saucers," how was "flying saucer" defined? In other words, what did this massive phalanx of scientists decide they were looking for, investigating or attempting to mathematically eliminate? There are two definitions given. One is a facetious one from Dr. Allen Hynek, who probably regrets today that he ever put it in writing. The other is an evasive and verbose definition, which makes *no reference to interplanetary vehicles,* either of terrestrial or non-terrestrial origin.

The whole object of the report, on which Secretary Quarles belief is based is not sensibly defined, and yet panels of unnamed scientists and mathematicians fed copiously at the public trough while they mathematically eliminated something they had not defined. I cordially invite these same panels of mathematicians to eliminate the "Amoeba" with the Chi Square test.

Then Secretary Quarles, talking about something his own expensive experts had been seemingly unable to define logically, could only dare to say "we believe," that no objects such as these undefined flying saucers had overflown the United States. People must realize that there is likely a difference between what we "believe" on the "basis of this study" and what is *known*. The end result of Project Bluebook, this monstrous catalog of misdirected scientific labors, is a "belief," not positive knowledge.

Then Mr. Quarles winds up with this "I feel certain." Does the public who is paying for these expensive shenanigans really care what he "feels certain" about? What does he *know*, that is the question! If he knows nothing, and these faceless panels of scientific gentlemen share his ignorance, let him say so. But the thing Mr. Quarles "feels certain" about is something covered by a nice big "if." "If more complete observational data had been available" —

If more complete observational data had been available it undoubtedly would have suffered the same fate as the rest of the data, namely, been subjected to mathematical processes of a rigorous nature whose sole purpose was to prove the non-existence of something the best scientific brains in the United States could not define, other than facetiously.

This obsession with mathematics and IBM machines, which is crowding out *original thought* will *not* in my view, permit the proper conclusions. Statistics can be made to prove anything, as any insurance company or public utility going after a rate increase will testify.

The thing that I feel certain of, is that if Mr. Quarles had personally directed and driven forward UFO investigation in the manner that its importance clearly warrants, panels of scientists would have been forced into original investigative fields, instead of using this government project as an iron lung to keep their incomes alive.

The attitudes of the air forces can only be explained in the light of an emergency. If information is in their possession which indicates malevolence on the part of beings or craft which we cannot control, then the Air Force attitude is justifiable. I believe this to be so, and believe that it is the only explanation outside of wanton negligence that will fill the bill.

On this same question, advice came from unseen beings, with Andolo the communicator in this instance: "I would add warnings against making the mistake of feeling your air force is committing some kind of crime against the people. Others have taken this attitude, despite adequate evidence to the contrary. To begin with, the air force cannot damage, harm or interfere with our craft. They are of a different order of matter to yours and indestructible by human agency. The reason that they are firing at craft when sighted endeavoring to bring them down is that they know they are being attacked. They also know the broad truth of what we have told you and what you have published. Therefore it would be in the interest and service of the people to keep the acrimony heaped on the air force to a minimum. They are doing their best in a difficult and oftentimes impossible situation, and their enemies, as you know, are also yours. Join with them as far as you prudently can in the efforts they are making to offset these activities of the dark ones. They cannot damage us, so there is no reason to become concerned about their shooting at us. It does not worry us."

There appear to be many germs of truth in this communication. I have seen many people actually become livid when they hear about Air Force jets shooting at UFO's. This is because to us, the worst welcome you can give anyone is to come out shooting when they visit you. The only likely reason that our planes are going out shooting is that they have themselves been shot at.

It is certainly not out of spite, nor is it mere unreasoning arrogance on the part of Air Force commanders. It is a retaliatory and a precautionary measure.

The concept of different orders of matter, and the immunity of the etheric forces to any attempted attacks by us would indicate that we are actually getting steamed up over something that does not worry these etheric forces in the least. On the contrary, the clear inference in this last communication is that the Air Force *has* to shoot at some of the UFO, and *they* (the etherics) understand this. There is no ill-will revealed in this communication, any more than we would be angered by a small boy trying to squirt us with a water pistol through bullet-proof glass.

According to some UFOlogists and students of the phenomenon it is this nasty welcome that prevents the craft landing here. This information from Andolo would seem to explain much of this, and our reason can do the rest. It might also turn out that the craft of the etheric forces are not intended for the purpose of landing on the earth at all, since it is to them a different vibration. It is my view that something in the nature of a three-cornered war is going on in the heavens around us, and that our airmen are participants in it on frequent occasions, battling against strange creations and perhaps still stranger creatures.

What information is known publicly about air force plane losses? Not much, but it is enough to be significant when viewed in the light of these Ashtar communications. Brigadier General Joseph Caldara, Director of Flight Safety Research for the U.S.A.F. issued figures for the year 1955 as follows:

There were 1664 major accidents to U.S. aircraft. In equivalent terms, eighteen squadrons of jet fighters (794 planes) and five squadrons of bombardment aircraft (75 planes) were lost. Over twenty percent of these losses were from 'undetermined causes'. This represents more than one hundred and fifty jet fighters lost from undetermined causes in the year 1955, or *three per week*. One might even ask "where's the war?" so large does this attrition seem to be.

If we add to this the equally strange disappearances mentioned by Mr. Harold Wilkins as having taken place in and around England, we have quite a total and quite a sombre picture.

It seems both obvious and logical from this information and from the other instances quoted in this book that the Air Force has its hands full. It does not one bit of good to get hysterical over their unwillingness to reveal what may be terrifying and to the mass of the people, indigestible details. I venture to say that the sudden introduction of all the known facts to the people, whether conversant with the UFO or not, would result in a high degree of panic being produced.

If one develops the capacity to read in between and behind published reports, one will find the truth. Here is a classic instance, from the Fullerton, Calif. "Tribune" of July 26th, 1956.

"Honolulu, T.H. (OCNS). The United States Navy will not publicly admit that it believes in flying saucers, but it has officially ordered combat-ready pilots to 'shoot-to-kill' if saucers are encountered, OCNS has learned.

"The information was first learned when Navy pilots navigating Trans-Pacific routes from the United States to Hawaii were ordered in a briefing session to engage and identify 'any unidentified flying objects'.

"If the UFO's (saucers) appear hostile, the briefing officer told pilots of Los Alamitos Naval Air Station reserve squadron VP771, they are to be engaged in combat. In Honolulu, members of the squadron talked over the unique orders. It was found that the orders are not unusual. They are a standard command issued to pilots on the trans-

Pacific hop. The conversation brought out other information, too. OCNS learned that, although the Air Force had publicly stated that it does not believe in the existence of saucers, extensive operational procedures, including forms of combat have been devised by various defense commands. The Hawaiian air defense command reportedly is on the alert at the present time, and if a UFO is sighted, pilots throughout the islands, armed with various and new-type weapons, will be scrambled into the air ready to fight. 'It's gotten so we wouldnt dare to say we've seen a UFO' a Navy commander told OCNS. 'If we did, every pilot in the Pacific would be ordered up. It would be pretty embarrassing if all we'd seen was a sunspot on the windshield.'

"However, the existence of the saucers themselves seems to be a generally accepted theory amongst the Navy fliers encounted here. 'I believe there are such things,' one pilot said, 'but I think that Washington might be wrong in their shoot to kill orders. The fact that saucers are in our atmosphere doesn't mean to me there's any pending invasion — which is what Washington seems to believe. And if there were an invasion, we'd do a lot better if we sent out a flight of priests and ministers, rather than a bunch of rockets and machine gun bullets. If anybody who could conceive a saucer wanted to invade us, there's no sense fighting them. They've got us licked from the start.' Operational procedures for a UFO scramble apparently are highly classified. Most officers refused to discuss the Pentagon's plans or modes of saucer combat. However, it was learned that a concrete plan of action does exist, covering all types of saucer sightings. The plans reportedly can be swung into action within seconds. The saucer sightings, however, have grown few and far between of late.

"Some of the Navy pilots here readily admit they have seen objects they believed were saucers. But none said they had reported them, either by radio at the time or by operational service reports when they landed. Reasons given for not reporting the alleged sightings ranged from 'possible ridicule' to a reluctance to put 'every pilot in the Pacific to work twenty-four hours a day for the next six months.'

"One officer pointed out that he felt if he reported a UFO he sighted last June 29, his very fitness to pilot a plane might be subject to question by Navy Brass. The pilot said that most officers he knew were of the same opinion. They may sight saucers, but they'll never report them, he said, unless the saucer should damage their ship through an act of aggression. 'Even then' he said, 'it would probably be better if you said you ran into a mountain. You'd sure have a lot less trouble.' The only solution to the problem, apparently has been foreseen by Navy officials. The only time a pilot would be proclaimed a hero for a saucer sighting, would be if he shot the ship down and brought back tangible evidence of his kill. This, perhaps is the major reason for the 'shoot to kill' orders currently being issued to Navy pilots crossing the Pacific. 'How do we know our bullets will work on a UFO?' a Navy pilot asked.

"Then the flier concluded, 'and if we do shoot, that's asking them to shoot back. And we don't know what they're going to shoot at us'."

This self explanatory news report contains some hidden gems for the UFOlogists. Chief of these is that the actual orders given by the briefing officer were that the UFO were to be engaged *if they appeared hostile*. This indicates that the authorities are already aware that not all are hostile.

The entire Hawaiian area is looked upon as a very bad area, psychically speaking. Sensitive persons with whom I am friendly are unanimous in this impression. It was also in Hawaii that the pilot was alleged to have caught a glimpse of the occupants of one of these UFO, as mentioned in Major Keyhoe's "Flying Saucer Conspiracy." The sight of the being apparently unhinged the mind of the pilot.

This newspaper story reveals that there are "UFO scramble procedures" and the first squeak of a warning turns the entire Pacific area into a hive of activity. This sounds more like the businesslike U. S. Navy with which we are familiar. They are doing what is, in the light of their clear duty, the very best they can.

The "shoot to kill" order also has reason behind it. It indicates that attacks on airplanes have needled the defense service into striking back, despite their patent knowledge of a *dual* manifestation. This fact is revealed in the briefing orders mentioned above, i.e. "if hostile."

The pilots might well be afraid of what the UFO might do to them, and wonder about the power of their bullets to bring down a UFO. It would be comparable to trying to riddle an actor you didn't like by shooting him off the TV screen. Most of the UFO would seem to have no more tangibility than a light image or projection, which some of them may be. In the case of the near-physicals I believe come from our own moon, I hold the view that it might be possible to shoot these down, even if unlikely.

The power of the UFO to retaliate to any unprovoked assault might well be shocking. Some of these eerie visitants unquestionably use spiritual or metaphysical methods of self protection, as indicated in the November 8, 1954 sighting at Monza, Italy.

In this most interesting case, brilliant illumination at 10:30 p.m. on a sports field drew the attention of a passing cyclist. A crowd gathered, then some of the spectators saw "figures in white pants, grey jerkins and transparent helmets." The glare from the ship was intense, illuminating two small entities, one of whom had a black face with a sort of trunk. (The reader is refered to the illustration ("A Spaceman?" in this book.) The crowd threw bricks, but they did not clang on metal, rather did they give the impression of landing on something soft.

Emitting a blinding *silver* light from its conning tower or cabin, the saucer continued to sit there until the crowd broke down the gates and surged into the field. The entities retreated to the disc, but not before one of the crowd set a boxer dog on to them. The dog immediately turned on its own master and bit him. This indicates a knowledge of self protection possessed by these entities that will have great meaning for students of occultism and metaphysics. It is inexplicable to orthodox science.

If this process could be repeated with cannon shells or bullets launched against similarly knowledgeable entities, who do not necessarily have to be "good" entities, our pilots might well finish up shooting themselves down as their bullets return to their own source in this manner.

The wisdom of the pilot who suggested sending out battalions of priests and ministers against the UFO rather than squadrons of planes, is well founded in that the bulk of the UFO appear to be of spirit or spiritual origin and nature.

Whether or not the average gentleman of the cloth would be in any position to deal with them however, is something that cannot be answered at present. The UFO are, in my view, more certainly the bailiwick of the "sky pilots" than of air force pilots, but it is very doubtful if the former are as well skilled and knowledgeable in their field as the latter are in theirs.

It should be the duty of the chaplains to give to our airmen the necessary spiritual armament they need when they are sent against what is essentially a spirit and spiritual manifestation. I see in this field great scope for all the noble qualities of the chaplain corps. Certainly no stone should be left unturned to see that our pilots have the best we can provide, but there are without doubt chinks in the spiritual armor of the men who go out to trade manoeuvres and sometimes fire with these weird entities from space.

Until a greater understanding of these things is obtained, we can all aid our airmen by prayer. And we can aid our air forces, by extending to them a more tolerant attitude in the light of the fact that they are trying to deal with hostile entities whose purposes, nature and motives they cannot begin to understand.

By making a conscientious, individual effort to understand more of the nature of things spiritual, every citizen can do his part to prepare his government for what would seem to be a very grave and portentous period. Each citizen can do himself a favor by developing a soundly functioning discrimination, that will permit him to classify trash as trash, whether it emanates from a government office or from an unseen entity grinding an unseen axe.

CHAPTER TEN

THE PHOTOGRAPHS

"Knock and it shall be opened to you." — Jesus.

The weakest feature of "contact with the UFO" stories is the lack of evidence, at least in the eyes of persons of orthodox viewpoint. Aside from the photographs of George Adamski, there is virtually no evidence that has been obtained *on purpose* that is in the public domain.

To a person like myself, whose essential concept of the UFO is that they are predominantly non-physical objects, beings or creatures, the need for actual evidence is not great. The reason for this is that after a time, even a passing acquaintance with the super-physical shows that the evidence of its existence is something largely within the individual himself. The standards and nature of the proofs are, in other words, different to those of the physical plane and must be approached differently.

Despite this, I became acutely conscious of the need for evidence amongst those first entering the subject, especially people of some intellectual power. For this reason, Jim Woods and myself decided to devote as much spare time, effort and money as we possibly could to gathering this evidence. We had no fundamental need for it yourselves, but we could clearly see the need of others. The whole project became an intriguing blend of original research, invisible living things and service to others.

Most of the persons involved in UFO contacts are essentially decent and innocuous persons. Most of them are sincere and well-meaning, although there are others who have been duped by invisibles and those who have become tools of some of these forces.

It is because most of them are harmless, shy and sensitive that they have been roughly handled by the official scientists and their stories mocked by the newspapers.

The information we are releasing in this book, and especially in this chapter is not harmless. It is going to do vast damage to the barriers erected against man's spiritual progress by official, orthodox thinking. In particular, the conditions for life in places other than this earth drawn up by official science can now be shown to be incorrect, and far from exhaustive. Proper scientific effort must now be devoted to classification and understanding of invisible but real creatures in our own atmosphere. And "invisible but real" leads science to the portal of spiritual matters.

The photographs in this book are but the first stumbling steps along the way. They have had the effect of impressing both Jim and myself with the paucity of our own knowledge, and humbling our all too human tendency to think that we know something.

Like the overwhelming majority of the earth's inhabitants, we are ignoramuses in these matters we have been investigating which are, we believe, an important key to human enlightenment. They are of the spirit, and not of the flesh. All that we have done has been based on man's glory as a spiritual being.

These published photographs are part of a large collection. We have been able to reproduce only those which would suitably respond to the half-tone process used for book reproduction. Scientists familiar with infrared photography will note the power of some of these radiations, and will be forced then to find out the source of such radiations. Military scientists will have to assess the potential effect on rocket guidance systems of such radiation, especially if it extends down from the infrared band of radiation into the frequencies they are using for these guidance systems.

Two alternatives confront these gentlemen. They can attempt to prove that the photographs have been faked, in which endeavor they will not be sucessful, or they can get back to their drawing boards and start thinking hard and originally about invisible intelligences that radiate. We take this opportunity of warning again those who will assail our efforts as fakery that there are trump cards in our possession of which they had best beware.

In the past year or more, we have made many trips to the Mojave Desert and other places in California in search of the elusive UFO. Side by side with these trips, we have both earnestly studied various aspects of esoteric teaching and superphysical science. We have practiced some of these things and obtained results that speak for themselves.

If at times the passage of this story of our photographic efforts seems to have holes in it, it is because there are some things of which we may not yet write.

In the early stages of our investigations and our experiments we discovered that solitude and peace such as that found in the desert are almost essential to success and results. We have not yet reached the point where we are able to demonstrate these things before audiences of skeptics, although I believe it is not far distant. Because we have in all instances been alone on the desert, we have covered these happenings with the affidavits contained in the appendix of the book. We have sworn to the truth of what we have done and experienced, and know enough of these matters to appreciate that affidavits may not be lightly made

To begin with, we state without reservation that it is possible for any person with a 35 mm camera and the necessary filters to photograph "flying saucers" over the Mojave Desert at any time of the day. The person will not be able to *see* the UFO, but when he develops and prints his film, if he uses the right film, he will quite possibly find that the skies in that region contain many of these constructs and forms which were optically invisible at the time the photograph was taken.

These are not Menzelian mirages, film flaws or phenomena that astrophysicists can blow down. They are real craft, or creatures, which hover or float constantly above the desert and perhaps in many other places. They appear to *belong* there.

The first and most essential "wrinkle" we learned was the use of infrared film, which permits photography of *invisible light or radiation*. We obtained this information *from Ashtar,* whose early suggestion it was, before I terminated contact with him many months ago.

These pictures which any person may take, often require interpretation, for many of the objects may be very small, and the others, so large and close in the film emulsion that they will escape the attention of the untrained viewer. The craft, being invisible, are not apparent to the human eye, but from their lower rim or flange radiation is emitted which affects the emulsion of the standard infrared film (Kodak I R 135) obtainable at most photographic stores. These rims and often other portions of the ethereal objects, show up as white semi-ellipses against the black background of the standard infared exposure, when it is printed.

To ensure that *invisible light only* is recorded, the photographer should employ an 87, 88A or 87C Kodak Wratten filter over the lens, although a 25 or 29 filter will also work even if the light passed is not strictly invisible.

For experimental purposes, it is not necessary to have any conditions other than a perfectly cloudless day. Take a few rolls of infrared film, and shoot the horizon in overlapping segments through the full 360 degrees. It is possible that in the Mojave Desert area, strange objects will appear in your pictures, which should be examined minutely. You may find, for example, a craft close to the ground very near you. You may catch an invisible flying animal or a creature. Try it and see, and you will find for yourself that we speak the truth in these things.

In obtaining pictures of these things at close range, and *on purpose,* different techniques are required in which the basic acceptance of certain occult teachings plays an indispensable role.

The immediate question is raised of how to aim your camera if you cannot see the objects. There are two ways in which this may be done. The first is the use of some device which will reduce or convert the infrared radiation to visible indications. The standard Army "Sniperscope" or infrared telescope will do the trick, although it is clumsy and rather insensitive to vibrations of the order we are dealing with in this case.

The second alternative is to use and develop one's own God-given faculties which are lying dormant in us and not depend on instruments other men have made. Having tried both methods, we can state that both have their value, and each is appropriate at its own time. The second method however, is the best, and produced all the pictures in this book.

Constant studying of the UFO phenomena from all angles, together with the new evaluations forced upon a person by the study of elementary spiritual science, produces an *awareness* of the presence of invisible beings, creatures, or craft in the immediate vicinity.

First, one becomes conscious that such things do exist, in the universe and then an awareness of their *actual presence* follows. It is this awareness that is the first point of breakthrough for the spiritually scientific investigator. All this, however, is getting ahead of the story.

Not being wealthy men, we had to commence our expeditions with borrowed equipment. Borrowed cameras became the bane of our lives, and so did rented cameras. Not only were we almost always forced to use inferior equipment because of this, but we never had any camera long enough to become accustomed to it and obtain consistently good exposures.

There are other reasons for using one special camera consistently in this work, which will be dealt with more fully in a separate book now being prepared dealing with UFO photography.

A business associate of mine came to our rescue when he perceived our difficulties and offered me the unlimited use of his German Leica. I was then able to purchase glass filters for this camera, and it was with it that the "Amoeba" pictures, the "Peekers" and "Adamski's Patio" shots were all taken, as well as many others.

The standard infrared film soon proved to be less than adequate for most of our work, and we ventured into high speed infrared film. This highly sensitive material comes in one hundred foot rolls and has to be spooled into cassettes by the investigator. After a few weeks of constantly fogged film, we also discovered that infrared light penetrates these cassettes and that film and camera must always be in total darkness during loading and unloading. The cassettes must be kept in cans when loaded, and only taken out of the cans in total darkness.

Between the loading and spooling of this film, and the difficulties involved in putting high speed infrared movie film into fifty foot magazines from one hundred foot rolls, we had many less than joyful experiences. For months we carried on this sweaty and exhausting process in my pantry, until finances permitted better facilities. Despite the difficulties, we did get the results.

All these things took place over a period of many months, but as our knowledge expanded, so were we able to learn more lessons that we are beginning to apply today. Bit by bit we eliminated the various problems of the photographic equipment, films and other apparatus until finally we were completely free of these and able to concentrate on the business end of UFO photography.

We invested in an infra-red telescope, in the hope that it would be the answer to our prayers, and in fact I did photograph one UFO from my own front porch using it. However, it was shortly after this that we really began to break through and its further use became unnecessary.

Constant attention to the subject, and concentration, began to produce an extension of perception in both of us. We both began to find that we could "see" portions of the auric emanations around other humans, animals and plants. This perception is heightened by learning how to de-focus the vision a little, and it was through the use of this rudimentary "third eye" that the pictures in this book were taken.

This improved perception, when combined with the awareness of the presence of invisibles, equips the investigator to take these strange pictures. This awareness has reached the point where it is not uncommon for either of us to awaken and sit bolt upright in the middle of the desert night, to find one of these characteristic vibrations emanating from invisible craft or beings above us or near us. Normally, we both require considerable stirring from a sleeping bag on a cold desert night.

Our big break came on Sunday August 25, 1957, around 7 a.m. We had camped at a site we have christened "Uncle Frank's Rock" in honor of Dr. Franklin Thomas of the New Age Publishing Co. We had been up since dawn trying to gain some "third eye" impressions of the presence of UFO. We had just fixed breakfast and had started to eat when I suddenly became *aware* of something above us. I said to Jim as I dropped my plate and kicked over my coffee: "There's something up there, look." Defocussing the vision in the manner that is necessary to see beyond the physical (you must look beyond it to see beyond it) I could see an extremely intense vibration which seemed to me to be like the flapping of wings.

I heard nothing, and Jim, who was also trying to "see" the object said, "I'm hanged if I can see anything."

Nothing daunted, I grabbed the Leica, which was fitted with an 87 filter (passes invisible light only) and opened the lens to its maximum f3.5 setting. With the shutter set at 1/30 second I began shooting.

The first picture is now history. It is the "Amoeba No.1" and shows a strange creature and some of its anatomical details. Above it, there appears to be another similar creature, although it is not clear to me whether or not this is part of the emanations from this strange creature. Note the rudimentary "eyes." On another occasion, we photographed four of these strange "bladders" flying above the desert floor, but the picture does not have enough contrast to reproduce herein. We have many others in the same category. They appear to be two dimensional, i.e. to have *no depth,* but this may be due to the particular way they register on the film. They are their own light sources, or to use the proper expression "self-illumined."

Amoeba No. 2, Amoeba No. 3 and Amoeba No. 4 were photographed in succession as the vibration moved away from its initial position directly overhead. I called out to Jim, "It's moving off to the south and getting lower," and kept on shooting. Two additional pictures make up this series but they do not have enough contrast to reproduce. There are seven successive pictures on this roll of film, the last one being the strange photograph labelled "The Peekers," which we will discuss shortly.

I have made an affidavit to the effect that I did not at any time see these objects or creatures optically, in the regular manner. Had I seen them clearly and optically, I would have been able to frame perfectly, instead of cutting off portions of them as I have done here. The reason for this is that when you raise your camera to squint through the viewfinder, the "third eye" simply drops out of use, and one must again defocus the vision to pick up the vibrations.

Clairvoyant persons, or persons with fully developed etheric sight, would not have this difficulty, but we were and are, only learners.

The "Peekers" shows two strange faces just caught by the camera at the end of the "Amoeba" series. The upper creature appears to be of the same order as the "Amoeba." The lower creature has a rather reptilian face with an extremely high forehead and strange bill or beak. We have another photograph taken near this location on another day which shows in outline what the body appended to this head looks like. The creature is apparently transparent even to infra-red, but its outline body as shown in the artist's impression from this second unreproducible photograph led to our christening it the "Flying Frankfurter."

The names borne by the photographs are those that became attached to them in our circle. Since we are dealing with unknown things, and they are not without their terrors, we have used names of a humorous nature in order to keep our balance.

It is interesting to observe here that the U. S. Air Force has referred nebulously from time to time to "Space Animals" in some reports, thereby arousing the ire of Dr. Menzel, Professor Emeritus of Astrophysics at Harvard and author of the book "Flying Saucers."

Dr. Menzel considered this Air Force suggestion (official U.S.A.F. release April 27, 1949) as facetious and out of place in such a document. Referring to extraterrestrial animals, Menzel further averred that one reliable report would be enough to justify all interplanetary saucer theories.

Dr. Menzel would doubtless consider us as not reliable, since no sheepskin has ever graced our walls. We are however, not going to permit Dr. Menzel to "pull the wool" over the eyes of others.

Extraterrestrial animals exist. We have photographed them and are well on the way to devising a repeatable method of doing so. The weight of this evidence will continue to mount, although it will likely be a long while before there is another picture like the "Amoeba."

Dr. Menzel's opinions further include the flat assertion that no one has the slightest evidence that extraterrestrial animals or beings exit.

Copies of our photographs are available to the Air Force, A.E.C. and even to Dr. Menzel if they are at all interested in conveying to the public or to their own minds evidence of what kind of life does exist, even if invisible, in our own atmosphere. The photograph labelled "Amoeba" No. 1 gives them such a creature to examine at their leisure.

Taking these photographs is an extremely hit and miss proposition for a person who cannot completely control his "third eye," but it is, in my opinion, the only way that photographs may be taken *on purpose* at the present time.

Each of these photographs has behind it a story, in some cases a story that the average person will not believe. All that we can do is tell our story, swear to its truth, and let the chips fall where they may. The passage of time will inevitably prove that we have spoken the truth, and contemporary acknowledgement of this is not expected by us.

In the case of the series of pictures taken in Mr. Adamski's patio, there is quite a story to tell.

For some time we had intended to pay Mr. Adamski a visit, and unfortunately chose a Sunday afternoon. When we arrived, we discovered that Mr. Adamski was due to deliver a talk to a group of interested people in just a few minutes. Not wishing to appear impolite and monopolize his time, we decided we would return at a later date when we could talk to Mr. Adamski alone.

We were standing in the patio of Mr. Adamski's home and had just turned to leave when I again became *aware* of the presence of one of these invisible things. Looking up, I "saw" a tremendous vibration around one of the nearby trees. I drew Jim's attention to it, and this time he, too, could "see" the vibration. The reader must understand clearly that we saw no form of any kind, only a vibratory effect.

I swung the Leica around, with the lens set at f3.5 and with an 87 filter over the lens. I made long exposures, (half second) as the film was the very slow infrared film, which accounts for the somewhat blurred effect of these pictures.

Adamski's Patio No. 1: An infrared photograph of the gap between two trees, from which the vibration seemed to be coming. Note that the tree on the right has just the suggestion of a form on top of it. Carefully note the shape of this tree, especially its top, and the large shadowed area at the bottom. Remember that infrared radiation shows up on the finished photograph as light or white areas, and absorbent objects or areas as black or dark sections.

Adamski's Patio No. 2: There now appears to be a small winged *being or figure* on top of the tree. We have come to call this the "Angel of Palomar." Note that the black area at the bottom of the tree has now moved upwards towards the top of the tree.

Adamski's Patio No 3: The winged being has been replaced by, or *has itself become,* the form of a flying saucer's edge. Beginning from the top of the tree, the lower flange or rim of the saucer can be seen, surmounted by the vertical, cylindrical section of the craft which usually has portholes or eyes in it. This in turn is surmounted by the dome-like top. Just the right-hand edge of the saucer is visible. Note the difference in position and extent of the black area of the tree, and also that radiating objects appear around the form of the saucer.

Adamski's Patio No. 4: The saucer has been replaced by a gigantic, heavily radiating object, the form and structure of which should be compared with a later photograph, "Frank's Freak." This form appears to be either a craft, or perhaps a monstrosity of some kind. Note the new position of the black absorbent area.

These four pictures were taken in rapid succession, and suggest *changing form* on the part of the objects photographed. Further photographs could have been taken, but unfortunately the shutter ribbons of the twenty-five year old Leica chose this moment to carry away, ending picture taking for the day.

Enlargements of Adamski's Patio No. 2 and No. 3 are included to show more detail.

Giant Rock Cloud: This picture of a "cloud" over Giant Rock was taken with infrared film on the Sunday morning of the 1957 Spacecraft Convention. The cloud was optically visible, of course, and serves to illustrate the theories about clouds and UFO held by me and described earlier in the book.

The Ether Ship: This photograph is named in honor of Meade Layne, who devised the term.

Usually on leaving the desert, we have a few exposures left in the cameras just in case we sight something or receive an impression on the way home. On the morning of November 16, 1957 we had packed up and were driving across the desert road when I became "aware" of something near us. I stopped the car, jumped out and picked up a very faint vibration in the air not too far from where I stood. I shot the last picture on the roll. I was so sure that I had something on it that I resolved to sacrifice the whole film for the sake of the final exposure. Knowing this vibration was very weak, I doubled the development time, and out popped this ethereal form in the sky. Note that it has a clearly defined, if ethereal-appearing top, and a very diffuse area of radiation beneath it. It is in more or less standard "saucer" form.

Frank's Freak: In December it became extremely cold on the Desert, and on one occasion I caught a chill and had to forego the desert for a spell. We had noted that coming home through Pomona, a certain type of UFO had a tendency to come down low over the road, although they defied our efforts at photography, with or without infrared film. Dr. Franklin Thomas evinced an interest in our work beyond that of a publisher of books, and we invited him to accompany us on a predawn visit to Pomona to see what we could produce.

We found a secluded spot above the freeway near some water tanks off Ganesha Boulevard. We gave Dr. Thomas a "Praktica" camera set at f1.9 and 1/25 of a second and told him to shoot, whether or not he saw the whites of their eyes, at anything that moved.

It was still dark when we dismounted from the car and Jim was still unloading equipment when we heard Dr. Thomas' camera "go off." Jim and I looked at each other and silently agreed that here was a "new chum" amusing himself at our expense. When asked he said that he thought he had seen a shadow overhead, that is, over *my head,* and had taken a picture. "Frank's Freak" shows a weird construct not dissimilar to that which thrust itself into the Adamski's Patio No. 4 picture.

In the early days of our activities, the method was simply to go into the desert and wait, full of hope, for some kind of manifestation, whereupon we would click shutters.

The inefficiencies of this method, especially since we have dealt all along with invisible forces are obvious. Since our conviction is now complete that the forces we are dealing with are of a spiritual or spiritistic nature, and we had felt this to be so from the beginning, we resorted to spiritual methods of attracting these UFO to our vicinity.

At this time, it is not appropriate or wise to state in full what was done. The answer is to be found in experimentation with body magnetism and orientation, which, when performed in a certain manner, attract certain of these denizens of the invisible realms.* All the remaining pictures were obtained by this method. I acted as a "pole" or generator, Jim as cameraman, with me carrying a camera as well in case I picked up something.

Force fields have been mentioned by ufologists, and scoffed at by astrophysicists. Here is proof that they do exist, and may no longer be laughed off or laughed at. They are very real, represent considerable power but remain for the present, invisible.

Force Field No.1: This shows the author standing atop "Ashtar's Rock," a vantage point we have often used for our experiments. The force shows itself as a field of radiation extending the full height of the negative. Jim Woods shot the picture.

Force Field No.2: This shows the author in the same position, five frames further along the same

*For hints on the distribution of body magnetism see —
"Man or Matter" by Ernest Lehrs.

film. The field is again present in the same way.

Force Field No. 3: This picture is adjacent to Force Field No. 2 on the film strip. Taken endwise, it shows the curve of the force field at its upper limit.

Force Field No. 4: This picture shows a heavy field with a strange protruding feather of radiation.

Force Field No. 5: Taken from the back door of my office in North Hollywood. The force or radiation shows as a solid white sheet going straight up from the roof.

Force Field No. 6: Shot at Giant Rock airport on the morning of May 3, 1958. We paid our first visit to George Van Tassel in over a year, and informed him that he had invisible visitors.

This photograph was taken in a clear sky while we were discussing UFO's with one of Mr. Van Tassel's associates. The camera was aimed by using etheric sight, and the force field containing a vehicle of some kind appears directly above Mr. Van Tassel's living quarters. Nothing was seen with the regular vision. F. 18 1/1000 second.

The Mojave Hubcap: We read with amusement Mr. Frank Edwards' recent article in "Fate" magazine on how to fake saucer photographs. One of his suggestions was that the faker throw hubcaps in the air. The radiation from the lower edge of this UFO, its highly absorbing center and seemingly neutral upper side may perhaps give us more hints on UFO propulsion than any hubcap. Nevertheless, we have named the photograph for Mr. Edwards, who has done excellent work in the UFO field. This picture lends itself to careful comparison with the "Vineland UFO" photo.

Being, Creature or Craft?: This shows a very strong shaft of radiation approximately twenty-five or thirty feet tall. One side is sharply delineated, the other quite diffuse. Some knowledgeable persons in the field of occultism have stated to me that this is a being. Others suggest it is the *projection* of a being. Others feel it is the emanation from a cylindrical craft standing on its end. This was one of the first photographs we obtained when we began experimenting with body orientation and gestures with mystical meaning. There is also a strong field across the lower right corner of this photograph. What do you think it is?

Two UFO and TJ: shows the author atop "Uncle Frank's Rock," during early experimentation with body magnetism. The two objects are semi-elliptical and are absorbing light or energy. Close study of this print in its original form prior to reproduction reveals many interesting things that didn't react with the emulsion of the film but which stirred it.

Flying Bush? This was shot by the author as Jim was setting up equipment. It appears in the photograph to be a bush, but examination reveals it to be quite possibly the *entrails of a UFO*, photographed as a sort of infrared x-ray. The form of the creature or craft surrounding the black "bush" can be seen in the original print. It is the standard flying saucer shape.

Auras No. 1 & 2: Two successive pictures taken of the author atop Ashtar's Rock by Jim Woods. No.1

shows a fairly strong and clearly delineated radiation field on the right hand side. No. 2 shows this same field in diminished strength, while a black aura or absorbent field or object of some kind has come in and moved the white field back to the right. The relative positions of the white field in both pictures can be measured against the hillock in the background.

A Spaceman? This was one of our earliest photographs, and shows what appears to be the helmeted head of a being in the left foreground. The morning was overcast, but the sun is reflected very strongly from this object as a white dot. The picture gives the impression of a small being in a form of breathing mask. At the time, I was photographing the saucer in the cloud in the center of the picture, and neither of us saw or were aware of this foreground entity. It remains to this day an enigma.

In December of 1957 I began to attempt to apply the same methods used in the desert in the area of my office in North Hollywood. Before long, various manifestations began to appear, and as best I could working alone, I attempted to photograph them.

Vineland UFO: Taken from the back door of my office, this object not only closely parallels the "Mojave Hubcap" in general characteristics, but gives a very fine impression of just how the various fields are radiated from these flying discs. This was optically invisible when photographed, and is a pure infrared picture.

The Big Shadow: Taken from the same place as "Vineland UFO." On this day I was very definitely aware of something in the area and I shot a half a roll of 35 mm film just as fast as I could, shooting in what appeared to be the general direction of the invisible object. The photographs before and after it on the roll are reproduced. There was no more than ten seconds between the three of them, so the "Big Shadow" came and went quite quickly. Note again how the object appears to wrap around and to some extent interpenetrate the parapet of the building.

The effect of this particular object was to leave a complete *void* on the negative of the film. We have subsequently photographed other manifestations involving these total voids, usually in the form of spheres or ovals. Nothing is seen with the regular optical vision, and this particular photograph was taken while employees of a nearby industrial plant were walking past a few feet away. The void probably caused by positive primary radiation which does not react with film emulsions.

UFO Ahoy: Shows the author photographed with a UFO on the Mojave Desert on March 8, 1958 just before type on this book was set. We had taken a number of photographs between dawn and 6 a.m. I was atop Uncle Frank's Rock and had run out of film when I became aware of something which seemed to be only three or four feet in front of my face. I shouted to Jim to "shoot" immediately, which he did with a Praktica FX2 using high speed infrared film. The object three feet in front of my face did not photograph, but one further away certainly did. The black object appears to be what is termed a flying saucer. It has a very black pin-point on top of it, indicating absorption of light, or a concentration of positive primary radiation. This would tie in with the theories of George Van Tassel, who claims that on top of these craft is a lens through which light is focussed to propel the craft. The white loops around the object are presumably caused by secondary but nevertheless invisible radiation from the object, and this photograph, together with the "Vineland UFO" and the "Mojave Hubcap" are probably the first photographs ever made of the force fields around these craft. They show how the fields alternate, or "push and pull" as they fulfil their propulsive task. This film was developed three and half minutes in D-11 developer, and exposed 1/50 of a second at F.-11. Careful inspection of the original print indicates that the small black or dark spot mid-point on a line between my eyes and the UFO may have been the "lens" of another closer craft, the one I "saw." The photograph is proof to us, although it may not be to anyone else, that what one "sees" is not necessarily reproducible photochemically.

UFO Shower: This photograph shows some two dozen UFO of the "Amoeba" type in flight above the Mojave Desert. This photograph is one of a series taken the same day. These creatures appear to contract and expand as they move through the air. They were not optically visible when photographed, but were picked up as pin points of light to the "third eye." The photochemical process reproduces them as *black* or dark objects, indicating that the polarity of "third eye" vision is the reverse of that of regular sight. They were photographed at F1.9 and 1/25 of a second at 5:30 a.m. on high speed infrared film. These creatures should become a legitimate part of space exploration, for they are real and they live in space.

UFO Shower No. 2: Shot by the author at 5 a.m. on April 26, 1958 in the same location under the same conditions.

The Author in a UFO Shower: This was shot by Jim Woods at the same time as the author was shooting the series containing UFO Shower No. 2. As another camera was used it indicates that the strange light beings simultaneously visible to our "third eye" vision, were in fact present.

It will be noted that almost all these objects, fields, creatures or craft have been photographed on the desert. The logical question that will follow from scientific persons will be "Why go to the desert?"

There are good reasons why we have invested so much time and money in these constant visits to the desert, and not the least of these is the factor of isolation. It is our opinion that many of these UFO do not particularly care for too much human society. Many of them may be in no way human, which would make their predilection for the wide open spaces a logical one.

Then there is the factor of silence. It seems to be universally accepted in occult circles that sound vibrations may repel certain types of these creatures, perhaps even be destructive to them. We have already advanced the view, although it is not an original view, that we are dwellers on the bottom of an ocean of air. Human experience informs us that certain audible impacts *under water* are very destructive to the life in that water. Hence the possibility of desert areas finding favor with the UFO because of this factor of sound. Recently, we have begun experimenting in *total silence* and expect to cover the results of this in another book.

Thirdly, there may be something in the vicinity of the desert that actually attracts certain UFO. They are more likely in our experience at least, to manifest in the desert than around cities, and in fact, can be seen flying at enormous heights and fantastic velocities above the desert on virtually any clear night.

Just as our own cities are gathering grounds for humans, so does the desert appear to be the UFO gathering ground. It is to these sites that scientific investigators must repair if they want positive experience of the phenomenon. It cannot be had through calculating machines and punched cards, nor can effective judgments or original thought be formulated without personally exposing oneself to the phenomena.

Another factor relative to the desert and the UFO which has a definite bearing on desert sightings is the factor of altitude and the static charges in the air. The clairvoyant faculty in any human being, no matter how rudimentary, will function more effectively under these conditions. The man who can see only one inch of the radiation around a human being in Los Angeles will see perhaps six or eight inches around the same human on the desert, and, if the light conditions are right, may even actually see the "double" at times completely withdrawn from the same human. All superphysical vibrations are more readily perceptible in the desert and the awareness of the presence of invisibles is sharpened by the silence, isolation and lack of extraneous vibrations.

By the same token, and provided proper techniques are employed, better photographs can usually be had of superphysical manifestations.

Southern California has for decades been a great gathering ground for enlightened people from all over the globe. Its special combinations of climate and soil attract persons interested in the inner life. There are many sound philosophical organizations in California, some of them close to centers of the UFO activity. All these things would tend to draw spirit and spiritual forces to this area more than to others where such factors do not prevail.

We have incorporated in our affidavits the statement that we have never at any time seen optically the creatures or craft that we have photographed. Despite the thousands of miles we have driven, and the enormous amounts of film we have exposed, we have never once ever *known* what was on any of the film prior to development. On developing the films, we have often been as surprised as any person viewing these pictures for the first time. They are truly and genuinely photographs not only of the invisible, but of the *unknown*. Nothing is more certain than that more and better photographs will follow, until finally a repeatable technique is established which will be of great assistance to so-called space exploration.

As a result of our investigations we find it a quaint conclusion that "flying saucers must be interplanetary" or that they are "interplanetary spacecraft." It may well be that *some* of them are, but it is much more likely in our opinion that many of them have indeed originated in, around or near the auric envelope of our own earth, and are no more Venusian, Martian or Saturnian than they are American.

The most potent result of our work, and one that must surely demand official attention as well as the attention and intelligent thought of every responsible person, is that invisible objects, creatures, craft or fields of force *infer invisible sources*. If we have done no more than to push the eye off the throne of the kingdom of reality, we will be most happy.

A great opportunity now beckons for the development of new film emulsions which will permit a full investigation not only of our own atmosphere and its often horrible denizens, but of the overlooked and often mocked optical theories of Johannes Goethe. We believe these theories to be correct, and have based our own work upon them to the maximum extent possible in a world organized according to Isaac Newton as far as optical equipment is concerned.

Will these splendid opportunities for investigation of the invisible but real worlds be seized upon in America? Or will they be passed by in favor of the numerous inanities connected with "space" research now being so prodigiously financed by Washington?

"Space" research should surely begin with the space around us and immediately above our heads. It is far from empty.

Full investigation along these lines we have stumblingly started cannot ultimately be avoided, and it is our opinion that the fruits of such investigations will do much to expand human consciousness, help to overcome fear of physical death, and generally advance man towards his full emancipation. Such work should take place in America. We hope it will.

Because these photographs were taken using unusual photographic materials and unorthodox methods the reader is urged to study the photographs in conjunction with Chapter Ten. Our work will not be understood unless this is done.

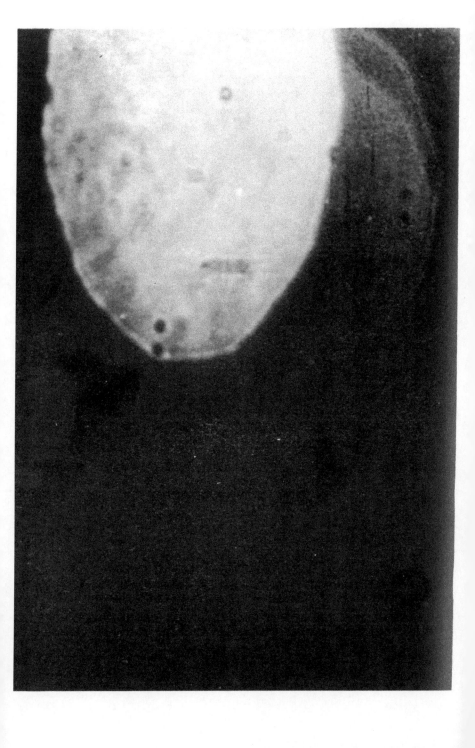

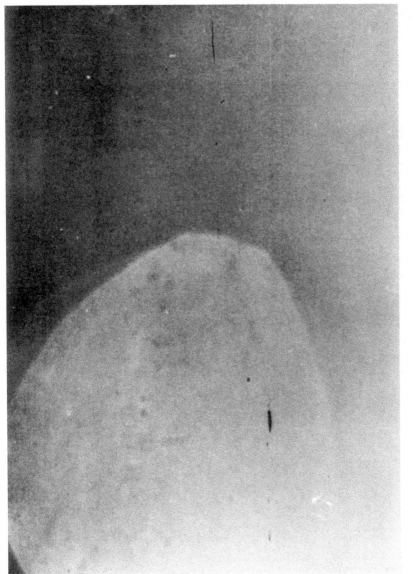

Amoeba No. 2.

Amoeba No. 3.

Amoeba No. 4.

The Peeker's

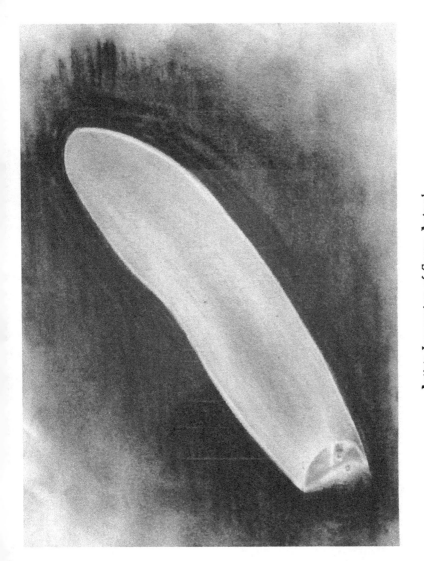

Artists Impression of Space Animal

Adamski's Patio No. 1.

Adamski's Patio No. 2.

Adamski's Patio No. 3.

Adamski's Patio No. 4.

In Adamksi's Patio No. 2. Enlarged.

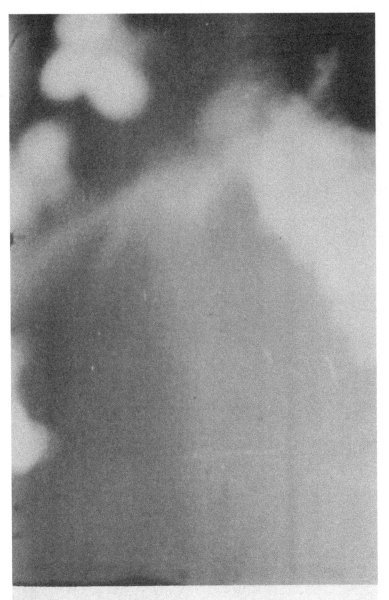

In Adamski's Patio No. 3. Enlarged.

Giant Rock Cloud

The Ether Ship

Frank's Freak

Force Field No. 1.

Force Field No. 2.

Force Field No. 3.

Force Field No. 4

Force Field No. 5.

Force Field No. 6.

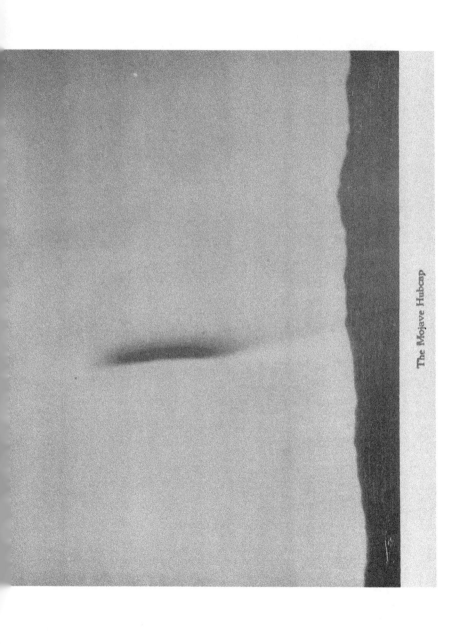

The Mojave Hubcap

Being, Creature or Craft?

Two UFO and T. J.

A Spaceman?

Flying Bush?

Aura No. 1.

Aura No. 2.

Three successive frames of film strip showing
both before and after "The Big Shadow."

The Big Shadow

UFO Ahoy

Vineland UFO

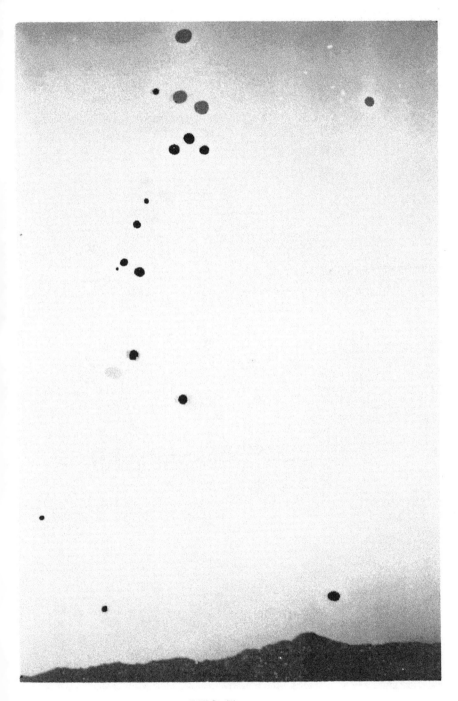
UFO Shower

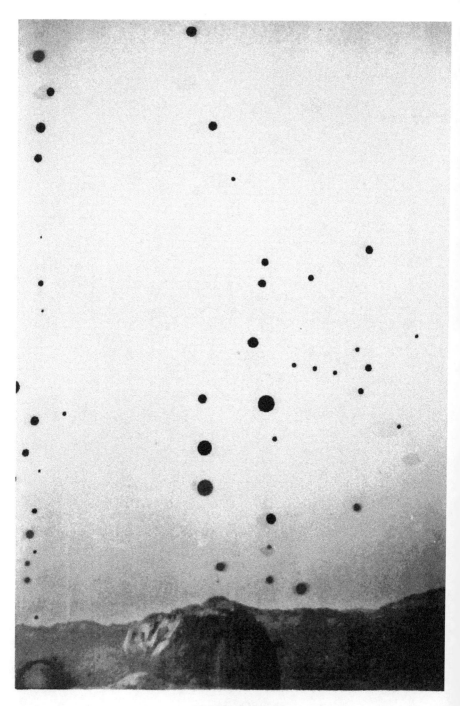

UFO Shower No. 2.

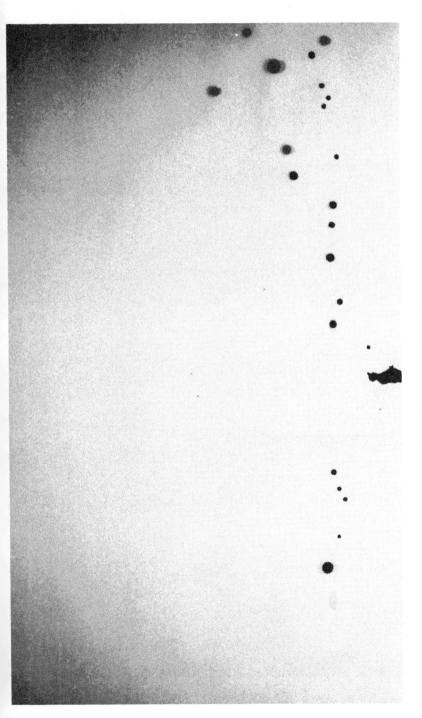

The Author in a UFO Shower

Force Field at 1 p.m. May 19th.

CHAPTER ELEVEN

SPACEMEN, VIBRATION AND LEVITY

"The things which are impossible with men are possible with God." — Jesus

Any attempt to classify the UFO on the orthodox aircraft system used by air forces, with silhouettes of various kinds, will produce a bewildering array of shapes and sizes. There are also the changing shapes which cannot possibly be classified by any silhouette and dimension method. They are already into the hundreds in their shapes and sizes and this fact alone has led to some interesting speculation regarding the craft by students of the superphysical.

The following question to Ashtar reveals something very interesting to all types of scientists who have been bamboozled by UFO's changing shapes.

Question: On the question of craft which seem to change shape, there is a school of thought which holds that these are the actual spacemen themselves, and that their bodies form the things we see. Could you enlighten us on this?

Ashtar: "This concept is essentially true. We are able to travel a great deal in our own etheric bodies. These are made of material so fine, that they are responsive to thought energy. Our bodies respond to thought with instantaneous exactness. However, we also have *craft* for the denser atmosphere of Shan. These are the Ventlas and other craft you have seen. It is a trial for a spaceman to come into the atmosphere of Shan. Its denseness, taint, and vibratory level are things they must endure in helping your people. They use the craft as you would a submarine, to travel in an unnatural, for them, medium. At times, it is necessary for them to leave their craft, in the same way as you might require skin divers to go under the surface of your seas to perform certain tasks. But they must have bases where they can return to their natural vibratory level and density, which prevails within the craft we use. Rest assured, however, that many of the strange and changing shapes that have been observed as unidentified flying objects are actually the spacemen themselves. Remember, we are not gravity bound, and not of the same order of matter as yourselves. I trust this clarifies the general query you have addressed to me."

To grasp this concept, one has to let the mind take a bold running jump at the unknown. Let the reader remember that many of these craft resemble nothing more than *diving bells*. Could it be that they are used in this way by beings of very fine composition who descend through what is no more than an ocean of air, to survey the life on the bottom of that ocean, namely, human beings and others?

Our oceans are not intended as a natural habitat for human beings. In order to function beneath the surface of them, we have to preserve the natural vibration of the earth's surface within submarines by having air available which the humans inside the submarine can breathe. Humans can and do venture into the depths with and without masks, although their ability to remain under the water for any period without air supply is *extremely limited.*

It may be that we have a similar situation with spacemen, namely, a limited ability to manifest in our vibration without their own natural vibratory conditions. It is possible that in the future, with the aid of suitable humans, some of these beings may be able to manifest here on earth virtually at will. At present, it does not seem that this is the case, although the rapidly increasing vibratory level of the earth may bring this about sooner than we think, and perhaps before we are ready for it.

In mentioning that their bodies "respond to thought power with instantaneous exactness." Ashtar may be taxing the credulity of the die-hard scientist and medical man. However, it has been shown in numerous experiments both in the United States and Europe that certain crystals can be altered in form and shape simply by the exertion of the proper thought power and control.

This is done with our own dense matter. What is illogical about beings of extremely fine matter, responsive to this kind of force, finding their vehicles (i.e. bodies) alterable by this agency? This fact is one that should be carefully borne in mind by all the investigators because of the bearing it has on judgments they may make of beings likely to appear on earth in the near future.

The dark forces from the astral regions of our own earth, must also have this power to change shape or form, in fact, it is a part of freely available writings on this subject. Let the person approaching or dealing with any beings who show up purporting to be from "space" learn *not to trust appearances!*

The whole idea of this ability to change shape and appearance may be a little indigestible to conventional scientists, but they had best get used to the idea, for it is likely to become increasingly a feature of our lives in the future. The ships, or craft may be themselves built or constructed by this same thought process, or, in other words be mind-constructs. It is more than possible that many of them are just that, and that they represent the body or vehicle of the occupying intelligence.

It is also more than possible that some of these craft, especially those from the etheric planes, are not intended for the purpose of landing on the earth at all, but may be no more than intervibrational vehicles for the occupying entity. After performing what function the entity may have to carry out, it may be that the construct is etherically disintegrated by the same thought process.

This whole concept of materialization is likely to become increasingly a subject for advanced thinking in the years to come. As a part of spiritistic phenomena it is already well known to investigators in that field. While we may assume that beings authorized to come here from what is purported to be the etheric level have the power to materialize, we must also remember that it is within the province of spirit beings from our own surrounding planes to materialize, and perhaps they are even now materializing *involuntarily* on occasions.

Due to my experiences with infrared photography, I am personally convinced that the veil is getting extremely thin between some of these planes. The condensation of water vapor used to conceal craft and described earlier in the book would itself indicate that on occasions the craft or beings do not have the power to *remain* invisible, and hence have to clothe themselves with a concealing cloud of water vapor to avoid detection. If our natural earthly vibration is rising, as all occult thought now seems to agree, it may well be that we are entering a period when some of these unseen worlds and their occupants, will be often and increasingly visible to human eyesight.

The question of materialization also leads us to the much discussed crashed saucers of New Mexico and Colorado. All Ufologists are aware that disc shaped craft, containing or purported to contain dead and scorched little men, are supposed to have crashed a few years back in the southwestern United States. This has been haggled over endlessly by various writers, some of whom may have been "had," and it has been denied by the Air Force. This latter fact need not confuse anyone seeking the truth. The saucers were alleged to have been crated and shipped to Wright-Patterson Air Force Base in Dayton, Ohio, for examination. The little men are purported to have been dissected by experts and found to be anatomically perfect.

I do not know whether these incidents are true or not, but if they are, it is my belief that a hoax has been perpetrated by satanic processes of materialization. This may be abracadabra to the Air Force and its satellite scientists, but it is likely to become a regular feature of this see-saw war in heaven as the dark forces endeavor to conceal the truth.

Earlier I quoted passages from Mr. H.C. Randall-Stevens' book "Atlantis to the Latter Days" in which a thumbnail sketch of the early history of the earth was clairaudiently dictated to Mr. Stevens. One of the great crimes of the deity known as Eranus, and now called "Satanaku" (Satan) was the materialization of "mortal sex bodies." In Mr. Randall-Stevens book, the scenes are described when these bodies were lying around after materialization with Satanaku *unable to put life in them.* The concept that "God is Life" would seem to be supported by this.

These bodies in the crashed saucers were dead, and the saucers supposed to be of strange material. I suggest that it is by no means impossible for this entire manifestation to have been materialized by this same dark deity. Who supposes that his skills in materialization would be any less after all these thousands of years? Acceptance of the origin and nature of the dark forces and their leader makes this entire happening appear in a different light.

Why anyone should conclude automatically that they had to be from Venus or Mars I cannot tell, although the possibility exists that they could have come from our own moon where we are fairly certain physical-type intelligences do exist.

The purpose of this particular manifestation, likely to be the first of many, is to delude. The purpose of it is to set everyone off the track and away from the *truth*. These aggressive forces are enemies of the truth and seek by every means in their power, including assorted materialization phenomena to conceal, withhold or distort the truth.

The finding of any such object by the Air Force would naturally lead them to think and conclude the thing that our own scientists want to conclude, namely, that the UFO are physical craft from other *planets*. We may expect that many such red herrings will be dragged across the path to the truth.

Those who have studied true magic and mysticism will know that the process of materialization is not carnival fakery, and in order that we be not deluded, made fools of, nor tricked into false assumptions, its reality and power must be made a part of formal UFO investigation. To ignore the matter is to invite disaster.

It would seem that certain of these craft and beings either dwell at or come down to a frequency range near our own where they can sometimes be seen, often photographed. There are many instances of UFO being recorded on conventional film in photographs taken of landscapes and other scenes.

The craft were not visible when the picture was taken, but gave off a sufficiently powerful vibration or reflection to affect the film emulsion. This effect is a likely one, in my opinion, with highly sensitive films such as Kodak's Tri-X especially where filters are used that cut off portions of the visible spectrum, but not the ultra-violet.

At these times, when the craft are not quite within normal optical range, they may be perceptible to radar, and to the consciousness of certain persons with extended vision. There are instances without number of radar echoes being returned from craft that cannot be optically seen, dating back to wartime when the U.S. Navy thought it was being "Kamikazed" by invisible objects. The radar showed flocks of "bogeys" converging on the carriers, which should have been in easy visual range. The objects were not seen either from the carrier's bridge, or from fighters sent up to intercept.

This pleasant exercise in the nature of reality may cause many orthodox theorists to writhe, but I am sure that the purpose of these manifestations is not to make us hang on to old theories, but rather, to compel us to *think out new ones.*

When near or close to our frequency, even though within the craft themselves the natural vibratory level of the occupying intelligence may be maintained, there would be a tendency for these craft to behave in accordance with the laws of this vibration of ours. Thus, sound has been reported on many occasions, together with vapours, angel's hair and other physical items. These things transpire when the craft operate in the physical vibratory level, or what is known to Ashtar as "gross matter."

As an example of this, sonic booms of an unexplained origin frequently take place all over the

United States, and in parts of the world where there are no supersonic airplanes. We know that some of the UFO have supersonic speed capability, or seem to have. In fact, if we take Ashtar's own word for it, they travel at the "speed of light and in multiples of it." Were any of these "booms" connected with the UFO? I asked.

Question: "Frequently there are sonic booms in this area caused by the passage of an aircraft through the 'sound barrier'. On many such occasions, these sonic booms occur when no supersonic aircraft are in the area, and in parts of the world where there are no such machines. Are these booms caused in any way by your craft or those of the satanic forces?"

Ashtar: "On the question of sonic disturbances in your atmosphere, frequently inexplicable by normal physical methods, I offer these advices. Our craft, and particularly those of the dark forces which are a lower order of matter than ours, frequently hover or drift above your surface, either visible or invisible, at high or low altitude depending upon the misson they are fulfilling. When they accelerate through this critical speed they, too, frequently break the sound barrier as you term it, producing the well known shock waves and booms when no other aircraft are in the vicinity. The velocities and accelerations of our craft are much greater than yours, but this does not always produce the violent boom. Rather does it depend upon how close to your frequency the machines are operating. This may help you clarify this question."

From this we may conclude that both friends and foes produce these sonic booms as they manoeuvre on their "missions." To this discourse we might perhaps add that it is quite possible that these etheric craft are disintegrated when their purpose is fulfilled, and also that there may be forms of combat going on in these invisible worlds which produce such sounds.

Elaborate theories have been formulated by many scientists, engineers and technicians to explain the ability of the beings to withstand the enormous gravitational pressures exerted by their observed manoeuvres. Ingenious men with inventive brains have produced many theories on these things, but they all have as their starting point the fact that there is a man, i.e. a physical human, in the craft.

Ashtar pointed out once that the physical body is a vehicle for the individual ego to permit it to function on the earth plane. If Ashtar is to be believed, the physical body is not required for functioning on any other planes in this solar system outside the earth-moon system. The likelihood is very great therefore, that we do not have a physical man of the same structure as ourselves to contend with, in theorizing about these craft.

Ashtar has stated that those of his particular faction at least are not subject to gravity in the same way we are. Therefore, g-force has no meaning for them. When a saucer comes to our frequency and takes on the appearance of gross matter, within it the spaceman remains at his own vibration, not subject to gravity in the same way as ourselves and he may quite possibly pirouette and zoom his craft in defiance of all Newton's laws, which seem to be more of the clutter of our science that needs to be overcome.

This is not to state that other theories are wrong, or that this particular theory applies to all craft. I believe that it may apply to some of them. We must beware of attempting to explain the UFO entirely with existing theories, rather than pursuing with original thought and *spiritual knowledge* new theories appropriate to the New Age we are entering.

Our approach to physics and mastery of the laws of life as they apply at our own level is not considered very intelligent by some of these invisible beings. At times, they have mentioned that conservatism turns easily into prejudice and prejudice into obstructionist activity when original thought seeks to open the way. Andolo characterized some of our activities in this way: "Physics upon your planet operates in accordance with laws drawn up to fit observed phenomena, usually with the purpose of explaining the phenomena. The result of this is that the true origin and starting point of many of your physical phenomena is not understood, although the end result is explained by the laws your physicists have devised for the purpose of explaining the phenomena without completely understanding the origin of them. To manufacture a physical structure capable of performing in similar way to our space vehicles certain physical concepts, time honored through centuries, will have to be discarded. Heat, light and sound are erroneous in concept on Shan. And yet they are the instrumentalities by which our craft are propelled.

Through other agencies it has already been pointed out that your science has become almost unbelievably complex and there is a reluctance amongst your scientists to accept any new concept *unless* it is complex. Believe us when we tell you, and through you, all your scientists, that complexity brings about its own downfall and that the greatest secrets of the universe are harnessed and employed by means so simple it would astound you.

Keep complexity to a minimum. Seek in scientific explanations the simple way to accomplish that which you desire. Consistent employment of this method will produce results rapidly and effectively."

And again, later, this personality stated in Sept. 1956. "Our interest has been attracted to the efforts currently being made to manufacture a craft of comparable performance and characteristics to ours as observed. However your airmen and scientists may try to do this, there are certain technological links that are missing. In order to secure these links, your science will have to revise many of its basic laws and concepts which are considered irrevocable."

That the U.S. government is in fact endeavouring to duplicate these craft of disc type was confirmed by Frank Edwards, the former Mutual commentator, at a 1957 Los Angeles lecture. Mr. Edwards told his audience that such a craft had in fact been constructed and tested in the summer of 1957 at Wright-Patterson Air Force Base. Of circular configuration, and fitted with some thirty-odd jets, this strange flat machine is controlled by a pilot who lies on his stomach under a canopy atop the disc.

Its first test was apparently unsuccessful, as the craft porpoised down the airfield without becoming airborne and finished up in a steaming heap at the end of the runway. And it is certainly a fact that enormous sums of money are being expended today on the conquest of gravity, although my own sources indicate that we are at last beginning to catch up with the real thinkers of our own age in this activity, and investigate and harness levity.

Levity is the natural force that exists as the polaric opposite of gravity. The universe is bi-polar in nature but most theories have sought to reconcile bi-polar electricity and magnetism with uni-polar gravity. This force is evident in nature in thousands of different ways, but outside of the Goethean school of scientific thought and other advanced centers, has not been investigated by conventional science until just recently.

In general terms, the levity field can be considered for rough purposes to extend spherically into space for a certain distance (perhaps 125,000 miles?) and its periphery represents its point of maximum force. Gravity may be considered to be a similar sphere with the center of the earth as its point of maximum force. Earth humans live close to the center of the earth, or point of maximum gravitational force, in bodies designed to withstand it. One consequence of this is that our consciousness has been *gravity bound.* Our observations are gravity bound unless and until we learn to look for the equal and opposite levity force in nature's book.

In our atmosphere, untold millions of tons of water

remain in suspension. What holds this water up there? Pressure, temperature or the wind system? No! The levity field! In the form of vapour the theory is that water's polarity is altered, so that in its new form it is repelled by the comparatively weak (at the earth's surface) levity field towards the point of maximum levity force, which is the periphery of the levity field out in space. As ice, water becomes gravity-polaric, i.e. heavy, and is *pushed* towards the center of the earth, or point of maximum force of the same polarity.

Apply heat to water, and its polarity is altered, and consequently it rises and falls in accordance with its polarity relative to these two opposing fields. Helium and hydrogen are substances that are strongly levity-polaric, and consequently we regard them as light to our gravity bound consciousness. In actual fact, these substances in balloons give a remarkable lesson in attraction and repulsion of the two fields, if we will but learn to read it.

Research into levity has already commenced in the United States, and perhaps those engaged in it would appreciate a hint from an unseen friend: "Levity and temperature are inseparably interlocked."

If our research can be made *pro-levitational* instead of *anti-gravitational*, we may find this to be one of the incredibly simple solutions to our difficulties purported to exist by Andolo.

In the fields of spacemen, vibrations and levity, there is much original work to be done.

CHAPTER TWELVE

RETROSPECTION AND SPECULATION

And when the thousand years are expired, Satan shall be loosed out of his prison. And shall go out to deceive the nations. — Revelation

When my first two booklets dealing with this subject were published, my apprehensions about them were considerable. I was not without fear concerning the acceptance they would find, and I fully expected that much of what I had to say would be extremely unpopular. My major reason for apprehension was my misapprehension that I was doing something new.

I soon discovered that what I had written was neither new nor first in its particular field. On the contrary, I was at least ten years behind another source, the Borderland Sciences Research Associates of San Diego, Calif. headed by Meade Layne, M.A.

In the perusals of Harold Wilkins' books, I had encountered brief references to Meade Layne's theories. At that particular time, however, these theories were so far in advance of my own thinking that I put them aside as so much hogwash. Beyond quotations used by Harold Wilkins and assorted unconnected references to Meade Layne and B.S.R.A.

elsewhere in saucer circles, I was not acquainted with the nature and substance of Mr. Layne's work. I had never read B.S.R.A. publications nor any of Mr. Layne's own writings.

It was with crashing surprise therefore, that I suddenly found myself the formulator of theories, and the recipient of extrasensorily obtained information that coincided almost exactly with Meade Layne's own now decade old writings on the subject. Far from facing the skeptical world alone, I simply stepped out on the salient driven into the "infallible phalanx" by Meade Layne and the B.S.R.A. To paraphrase Dr. Harlow Shapley, I was not alone.

Despite the fact that my extrasensorily obtained information substantially coincides with that obtained by Mr. Layne through the mediumship of Mark Probert, I still did not and will not join B. S. R. A. nor read their publications to any extent. By presenting Meade Layne with the first psychically obtained, completely independant corroboration of his work, I felt that a mutual service had been rendered. The key to any future correlation of independently obtained information or evidence lies for me in not getting any of the current B.S.R.A. theories down into my subconscious, from whence they might later be conjured and presented as my own. Hence I do not read B.S.R.A. material today.

As early as 1946, when information was very scanty, Mr. Layne obtained the basic theory of the UFO which has endured through the intervening years, and which my own information supports. In 1946, seven years before Dr. Donald Menzel of Harvard "shot Santa Claus," the whole UFO phenomenon was looked upon as being a sort of science fiction craze. Astrophysicists and astronomers in those days could load up a blunderbuss with astrophysical and astronomical facts and blow holes through flying saucer theories. Men like Meade Layne in those times ran the risk of having their sanity impugned for holding the views they did.

As a comparative Johnny-come-lately in this field, I feel it is most important to point out that credit for being either first or one of the first to formulate these theories rightfully belongs to Meade Layne and the B.S.R.A. The work of this organization deserves the support of every true New Age thinker.

It is also interesting that after he received my two initial booklets, "Spacemen, Friends and Foes," Mr. Layne interrogated the Yada (one of a panel of invisible but very much alive philosophers) through the mediumship of Mark Probert. Mr. Layne read the Ashtar communications to this being, who steadily confirmed and expanded them, without contradicting. At the end of a two hour period, the Yada stated: "Tell Mr. James that he talks too much, and too near the truth. He is in danger from the enemies of truth on both the physical and invisible planes."

Of course, there is no reason why the Yada should necessarily be right, but if he is invisible it may well be that he has had intimate personal experience of the things of which he speaks. His statement is an interesting support of the theory developed by myself from a source as invisible as the Yada.

The UFO today still *defy explanation* by orthodox official science. The question has become one of "who shot Doc. Menzel?" rather than one of that Harvard astrophysicist and his ilk "shooting Santa Claus," in their jealous desire to keep all sky phenomena confined to their own bailiwick. Dr. Menzel's theories, outside his explanations of sun dogs, refractions and reflections make odd reading today, and his book "Flying Saucers" is considered by most Ufologists to have been precipitate and ill-advised.

The lesson of the "haunted decade" is that some e.s.p. material is holding up much better than orthodox astrophysics and astronomy, in explaining the UFO. Great care should therefore precede the blanket condemnation of e.s.p. based theories, lest the person rendering such judgment suddenly finds himself aboard an intellectual roller coaster from which he cannot dismount. The UFO, whatever their origin and nature, are thrusting new concepts at us, concepts which practically dictate that we discard mechanistic thinking if we are to be prepared for the age into which our globe is being irresistibly propelled.

The superphysically obtained information has almost all come from persons who have studied the spiritual nature of man, and who recognize man as a spiritual organism. Certainly this is true of all really worthwhile material gained in this way. Unless and until a new branch of science becomes established or *recognized* which has as its cornerstone the fact that man *is* a spiritual assemblage and not a piece of fleshly machinery, official science will wallow and flounder as it endeavours to explain the UFO.

At the present time, the superphysically obtained advices represent a valuable guide in theorizing. However, because of the various factions involved in the UFO in the invisible of which there are at least two it is inevitable that tares will come in with the wheat. The wise process is a sifting process in which the cultivated discrimination sorts the wheat from tares. If trained scientific intelligences could be supplied with the required spiritual understanding, which is not to be confused with theological and sectarian dogma, we could in this way forge an invaluable instrument for the service of mankind.

We are moving into an age where we may expect various types of spirit and spiritual manifestations to occur, the UFO being only a preliminary. I further speculate that within a few years, the entire human race will be actually seeing things of a presently unknown and unsuspected nature. Some of them will be of the order of the "Amoeba." If we do not bend some effort to the establishment of a genuinely spiritual basis for our national way of life, we shall be caught with our intellectual pants down in this new age that is rushing upon us. In the words of General MacArthur: "There must be a great spiritual recrudescence".

In America, we cannot be "One nation under God," when half the nation is under Mammon.

Conventional science is preoccupied with weaponry as the highest form of endeavour, and this preoccupation with the destructive process indicates that our planetary sickness is of the spirit. And even the laws of conventional science and reason are violated by it. Science is a complex of theories, all intended to give proof of certain natural laws. Any theory, to survive, must be provable. The theory behind destructive scientific effort, is that it will provide security. The clear lesson of history is that weaponry has never provided security for any nation in history but only successively greater holocausts. It is therefore scientifically untenable that weaponry provides security, since all the proof is that the opposite is the case.

The human race will inevitably get exactly what it deserves, and it must be expected that enormous force will be required to administer to mankind the lesson that it needs. We certainly have this force. Perhaps official science is to provide the necessary janitorial services at our planetary house cleaning, pulverizing the old to make way for the new. We cannot penetrate these mysteries, but only speculate on them, especially as they relate to the UFO.

The reader might well ask now "How do the UFO fit into this?" To this logical question, there is an answer, although it must remain in the field of speculation for the present, until such time as everyone does finally agree that the UFO are here and want and mean something.

For some time I sought to fit the UFO into the world situation. From the time I first became aware of the fact that two kinds of UFO were involved I felt sure that the warring and hostile nations of this planet would somehow have invisible allies. I sought for the thread that would tie all these things together. The thread appeared when I learned that one Joseph Stalin had been the recipient not only of orthodox priestly training (which I previously knew) but of additional corrupting training in the art of black magic, and similar skills employed to satanic purposes. His spiritual alliances can hardly have been in question for this reason.

The thread becomes a cord, and the cord a cable binding this spiritual alliance when we realize that in Soviet Russia religion is regarded as an "opiate," and atheism is a civic virtue, and that the materialism of that country is epitomized in that Soviet youth organization known as the "Legion of the Godless."

Let the reader ask himself this question. If one be godless and boast about it, if one condemn the spiritual seeking of the humble as "opiate," which invisible faction would be drawn to such a person? The answer needs little imagination. It is obvious that such a person or nation would find on his side that deity whose stern resolve was to set at nought all the power of the Godhead, to strive with the Godhead for mastery of this earth. Satan.

The Bible describes Satan as "The Prince of the Powers of the Air." To an official scientist, this is religious jabberwocky.

But is there in the Western World, an intelligence officer who has not at some time been amazed by the technological advances of the Soviets in things pertaining to the *air?* Is there any person who can look at their achievments in the air, realize that forty years ago this nation was ninety-five percent illiterate and not be astounded? Can anybody seriously doubt that Russia, which entered the twentieth century with tools of the thirteenth has gained a remarkable ascendancy in the air? And all this has been done despite German devastation of their best food producing areas, and the slaughter of six million of their people by the Germans.

The answer does not lie, and clearly cannot lie as "Life" magazine so blandly asserts, in the capture of a handful of German experts. Any scientist knows that such achievements are the results of an organic and integrated scientific thinking, in which even second rate brains are appropriately used.

The West got the services of the top German rocketeers, Von Braun, Dornberger, Ley and Oberth and yet the Soviets vaulted into space research far ahead of us with their Sputniks one and two.

The superscientist *knows* that discoveries and research are boldly advanced by the *impression* of ideas from the invisible planes on suitable humans. In the case of the Soviets, discoveries and progress of a remarkable kind relative to the conquest of the *air* have probably been impressed from invisible sources friendly to the Soviet cause and having mastery of things of the air.

It is stupid to conclude here that whenever a Soviet engineer gets a problem he tunes in to the waveband of the Prince of the Powers of the Air and gets the answer. The process is not detected by anyone being impressed unless and until he becomes *conscious that such things can be done*.

It is altogether remarkable that every time the West produces a new bomber or fighter, the Soviets show up with something as good or better immediately afterwards. Either that, or they beat us to it. The concept that some of these abducted airplanes may be finding their way to the Soviets cannot be discounted. It may be as mysterious to the Soviets as it is to us, but it is possible that some of these jets which disappear land in Russia as mysteriously as the T-33 jet landed at Bolling Field not long ago, *without fuel or pilots.* It gives us something to think about.

We can, I believe, discount contact with these dark entities by the Soviet rank and file, but what about the Soviet leadership? The communist hierarchy can be considered today to be the vanguard of the anti-Christ. I resolved to ask my invisible communicator.

Question: What allegiances, if any, exist between the Soviets and the dark ones?

Ashtar: "The question is a difficult one to answer because it causes us difficulty in separating the Soviet people from their leaders. The particular type of Lemurian incarnates now ruling Russia are in direct allegiance with the dark ones through processes of black magic and other powers. The Russian people as you know them are no different from yourself and others. It is necessary therefore every step of the way to draw the line between the Soviet people and their leaders. The allegiance is most definitely there between the leadership and hierarchy of the Communists, and the dark forces against whom we war.

"Therefore if you use this communication draw a very solid and certain line between the Russians and their leaders."

The agitation against the atomic bomb by the Soviets has been ceaseless and virulent. A drum-fire of peaceful overtures has served as the means of drowning out the screams of their victims. Why has this agitation been so constant? If they themselves have the bomb, and harbor aggressive intentions, why would they seek to ban its use?

We are led directly to the Ashtar communications for the answer. Ashtar states that the dark forces cannot withstand the atomic explosion, because it annihilates them, shatters them and drives them down out of the air where they belong.* Is it not a fact that the Soviet Union has tested or fired only a small fraction of the number of bombs exploded by the United States? Why? And why have they almost all been of a much smaller size than those of the U.S.? Could it be that high directives from the Kremlin limit such activity in the interest of the invisible allies? Here indeed is adequate material for speculation.

Assuming that all atomic bombs were banned, as is now so ardently desired by Soviet Russia, what would be the result? The Western World, presently the repository of Christian culture, backsliding and all as it may be, would be outnumbered and outgunned. So vast is communist preponderance in conventional war-making assets (chiefly manpower) that defeat of the west would be well-nigh certain in any conflict.

The satanic forces in the invisible, who fear this atomic bomb as they fear nothing else, would then be

---

*"And the powers of the heavens shall be shaken."—Revelation.

left to go freely about their business, while their physical minions ruled the world which the satanic powers have so long sought to master for themselves.

The satanic powers, unable to leave the atmospheric or perhaps auric envelope of this earth, have no answer to this atomic bomb, the particular effect of which is deadly to them. Its present possession by East and West alike produces a Mexican stand-off, from which neither side dares to make the fatal move.

I speculate strongly that the atomic bomb has been delivered into mankind's hands that purposes other than his own be served, purposes and plans laid in a place and by a God whose plans do not go astray. In its possession and use, we will at all times be mere instrumentalities of that Power, regardless of what power we may attempt to assume.

It is but a short step from this speculation to another, that of missiles. The American effort in this field has been abortive, propaganda notwithstanding. Structural troubles have been most frustrating, and progress has not been in keeping with the purported technological leadership of the United States. I speculate that missiles in flight have been tampered with, and will continue to be tampered with, by these same satanic forces into whose workshop we are hurling our weapons.

Are these forces likely, if it is in their power to prevent it, to permit us to gain ascendancy over their physical allies, their earthly tools? I think not. Electronic equipment and guidance systems should be particularly vulnerable to the magnetic powers these orders of beings wield, and which in November of 1957, immobilized engines and extinguished headlights of cars in Texas. I have experienced the same force, differently applied, on the mechanism of my own car as previously related. This, together with the pictures I possess of tremendous fields radiated from an unseen source permit me no other conclusion.

Strange indeed are some of the things that have happened to American missiles. The "Navaho" ramjet missile, for example, after costing the U.S. taxpayers over $600,000,000 was cancelled because of repeated failures. On Aug. 12, 1957 just a month after the whole program was cancelled, the first successful test firing of the missile took place at Patrick Air Force Base. The missile went some 350 miles, without a hitch. Four previous firing attempts ended in failure because of *premature stoppage* of the two booster engines.

The problem was apparently whipped on the fifth try, *after* the development program was cancelled. Surely a strange coincidence. The Air Force, Navy and the Army must learn the patent lesson that certain UFO can and do STOP engines, and that they radiate unbelievable force in the infrared range of our spectrum.

There is evidence of other tampering with these things, such as the incident mentioned in "La Courrier Interplanetaire" of August 10, 1957, issue 31. The launching of the "Atlas" I.C.B.M. at Cape Canaveral, Fla. is mentioned, with the somewhat startling information that "a reporter belonging to a Jacksonville newspaper stated that just before the explosion, which he described as a brilliant orange, he saw what seemed to be a light or an object plunging towards the missile." The "Atlas" exploded in mid-air shortly after launching in this case.

No mention of what the Jacksonville reporter saw has ever appeared in the U.S. press, despite efforts on the part of one of my researchers to locate it. The report as reproduced above may be erroneous, or the incident suppressed. The reader must take his choice, since no one will hand him the truth in these matters on a silver platter.

It is with the power of retrospection that we gain a most interesting perspective on the future of the UFO, on which we are now speculating. Retrospection in a dispassionate manner reveals that the firmest explanations of the UFO yet obtained are those with a superphysical basis. While no one of these has the whole truth, official science is virtually unable today to offer an explanation on the basis of present knowledge that will stand comparison with the observed phenomena. The only answer they have for the many monster and little men sightings is coarse humor and crass reflections on the integrity and sanity of the witnesses. The superphysically gained information provides a place and a purpose for all these things. As they do exist, they must have such a place in the scheme of creation, whether or not official science has yet ordained what it shall be.

Official science cannot explain why it is that a jet can fly right through a UFO, but the superscientists can, even if they are unable to produce the externalized proof that is part of conventional science. They cannot produce the proof because it is a part of a different phase of science, in which the physical is rightly looked upon as the lower manifestation of man.

Official science cannot explain why it is that the UFO returns, or can return a radar echo but is optically invisible to pilots often flying right next to them. My own photographs show that life forms of one kind or another do exist which are optically invisible, and that they are frequently large and extremely weird in appearance.

The work of those of us who are investigating the UFO from the superphysical viewpoint can only go forward. If the conventional scientists do not bring themselves into phase with the spiritual manifestations and spiritistic forces impinging upon our planetary life, they will soon be little more than beaten men on a beaten path leading away from the truth.

Official science has conspired with governmental agencies to conceal things that it cannot explain. Virtually every investigator from the superphysical standpoint has made his findings and his work public, because it is the view of these persons that the truth will be found by seeking and that the more who seek the more humanity will learn.

The official scientists, under conveniently administered secrecy oaths, are able to conceal their intellectual nudity from the public gaze. They are protected from saying "I don't know."

The investigator from the superphysical viewpoint, realizing his own fathomless ignorance of the spiritual and spiritistic subjects, is glad to confess his ignorance and parade the fact that he has been humbled by the breathtaking might of the glimpses he has caught of these new worlds.

This is the picture we get from retrospection. One science being rapidly outmoded by planetary conditions, being dragged, kicking and screaming, into the New Age. The other science, the new science, seeks to meet, and to master the new age and its mysteries, that all may bask in its spiritual luminescence. As the New Age advances, doctor and saint will find precious little to support the great argument "about it and about," and both "will go in by the same door" eventually.

When failure, frustration and disappointment come the way of the properly financed scientist, he is able to overcome his failure with the cry "back to the drawing board." If we are to avoid failure, frustration and disappointment in the new age, we must finance ourselves with the resources of the spirit, and our cry must be not "back to the drawing board," but "back to the prayer rug."

By increasing use of prayer and meditation, we shall gain understanding not only of the UFO in their myriad modifications, but of the Universe itself.

Our work has never depended upon money for its advancement, but upon prayer, meditation and understanding of what has already been shown to us.

The future of UFO research lies in the hands not of scientists, but of *spiritually awakened people,* whether they be from the regular scientific professions or from the ranks of the most lowly and humble workers. These great secrets are not for the gross, the arrogant and the vain, but for the enlightened, the awakened and the sincere. They are for those who serve mankind, and not for those who seek to sabotage the spiritual heritage of man. In the realm of spiritual science there is no place for those treading the path to self-aggrandizement.

The first step to take if we have this pressing desire to know about the UFO, is to learn about the true nature of man, and his divine origin. This is the fundamental key.

And a step that everyone can take immediately is to expand the consciousness to encompass the fact that there are indeed spiritual realms, peopled with beings and replete with civilizations as real on their levels as we are on ours. To us, in our present earthbound state, "They Live In The Sky."

APPENDIX

Affidavit
GENERAL

STATE OF CALIFORNIA,

County of __LOS ANGELES__ } ss.

__MR. PIERRE PERRY__

BEING FIRST DULY SWORN, deposes and says

As President of the Copper Mountain Mining Corporation of Arizona, I was on my way to inspect a mine in 1943, at a mountain location north of Prescott, Arizona. The mine is almost inaccessible.

Close by runs the River Agua Fria, and on a hot summer day around 5 p.m. I was leaving the camp for the mine with two other men. These were another prospector and a Mexican miner named Isadon Montoya from Marinette.

We were on horseback, fording the river. All at once, Montoya, who was in the lead, yelled "El Diablo, El Diablo".

We looked up, and overhead a most terrific drama was unfolding that lasted only a few minutes. A military plane was in sight, so were two large unidentified flying objects that looked like balloons (Montgolfiers) without baskets. They were luminous and bright as the sun. The U.F.O.'s stood still as if waiting for the plane to approach, then pounced towards it. At the same time, they projected a violent luminous ray that could be compared with the large beam of a lighthouse.

The air vibrated with a terrific explosion as the plane was struck and came down. We saw two pilots bailing out, but as their parachutes opened another fiery beam was projected from the U.F.O.'s. The chutes took fire and the two helpless men fell to the ground to be crushed to death. The two bodies were later found.

Meanwhile, the frightened Montoya was praying and crossing himself and repeating, "El Diablo, Senor....I have seen the same thing many times senor..."

Then, from the horizon, coming from the north at an unimaginable speed we saw another U.F.O. It joined the two above us, and together they shot like lightning to the south.

We turned back to notify the authorities, and tell them what we had witnessed. Somehow, they had already been alerted, and a truck and a jeep were on their way. We met them and guided them to where we had seen the wrecked plane fall. Parts were scattered all over the mountainside.

It was a deliberate assault on this military plane, but at that time there was no talk of spaceships like there is today. Later I saw a sketch made by an airman, and I recognized the same U.F.O.'s I had seen that day in Arizona.

I am a U.S. licensed pilot, and a member of the Aviation Club of Savoie-Aix-Les-Bains, France.

Pierre Perry

SUBSCRIBED and sworn to before me
July 18th, 1957
M B Fletcher
Notary Public in and for said County and State.
My Commission Expires Feb. 28, 1958
(NOTARIAL SEAL)

Affidavit
GENERAL

STATE OF Illinois
County of Edgar
} ss.

Mr. Eugene Metcalf

BEING FIRST DULY SWORN, deposes and says:

On March 9th,1955,at approxmately 5:50 P.M. I witnessed the "plane-napping" of a jet plane while standing in my backyard at Paris, Illinois. The plane was coming toward me from the southwest and was traveling in a northeasterly direction. As I stood watching this plane, an odd-looking craft came from behind the plane and just swallowed it. The U.F.O. had an opening that was in my line of vision, and through the opening it took the plane. After this, the U.F.O. hovered and pulsated and churned up and down. Then it seemed to whirl and lift upwards.
While going through these gyrations, vapor came from porthole-like openings around the bottom part. The plane and the U.F.O.were in perfect view, and stood out clearly against the sky.
The object was bright silver and I heard no noise.
The U.F.O. was very big and bell-shaped

Eugene Metcalf

SUBSCRIBED and sworn to before me this 2nd day of April, 19 57

(Mrs.) Charlotte Kramer
Notary Public in and for said County and State.

(NOTARIAL SEAL)

Affidavit
GENERAL

STATE OF CALIFORNIA,
County of **LOS ANGELES** } ss.

TREVOR JAMES

BEING FIRST DULY SWORN, deposes and says:

FOR OVER A YEAR I HAVE BEEN ACTIVELY ASSOCIATED WITH JAMES O. WOODS OF LOS ANGELES, CALIFORNIA IN PHOTOGRAPHIC RESEARCH AIMED AT ESTABLISHING A REPEATABLE METHOD OF PHOTOGRAPHING THE PHENOMENA CLASSIFIED UNDER THE GENERAL TERM "U.F.O."
THE ACCOUNTS OF OUR EXPERIENCES, OF THE METHODS EMPLOYED AND OF THE EQUIPMENT USED CONTAINED IN THIS BOOK ARE A TRUE AND ACCURATE RECORD OF THESE ACTIVITIES.
I HAVE AT NO TIME DURING THE TAKING OF ANY OF THESE PHOTOGRAPHS ACTUALLY SEEN ANY OF THE MANIFESTATIONS, CREATURES, CRAFT, FORMS OR FIELDS OF FORCE WITH THE REGULAR PHYSICAL EYESIGHT.
I MAKE THIS DEPOSITION OF MY OWN FREE WILL AND WITHOUT A MENTAL RESERVATION OF ANY KIND.

Trevor James

SUBSCRIBED and sworn to before me this _26th_ day of _April_, 19_58_

[signature]
Notary Public in and for said County and State.

(NOTARIAL SEAL)

My Commission Expires Oct. 25, 1959

AFFIDAVIT—GENERAL—WOLCOTTS FORM 292—REV. 12-50
59300

Affidavit
GENERAL

STATE OF CALIFORNIA,
County of **LOS ANGELES** } ss.

JAMES ORVILLE WOODS

BEING FIRST DULY SWORN, deposes and says:

FOR OVER A YEAR I HAVE BEEN ACTIVELY ASSOCIATED WITH THE AUTHOR OF THE BOOK "THEY LIVE IN THE SKY" IN PHOTOGRAPHIC RESEARCH AIMED AT ESTABLISHING A REPEATABLE METHOD OF PHOTOGRAPHING THESE PHENOMENA CLASSIFIED UNDER THE GENERAL TERM "U.F.O."

THE ACCOUNTS OF OUR EXPERIENCES, OF THE METHODS EMPLOYED AND OF THE EQUIPMENT USED CONTAINED IN THE BOOK ARE A TRUE AND ACCURATE RECORD OF THESE ACTIVITIES.

I HAVE AT NO TIME DURING THE TAKING OF ANY OF THESE PHOTOGRAPHS ACTUALLY SEEN ANY OF THE MANIFESTATIONS, CREATURES, CRAFT, FORMS OR FIELDS OF FORCE WITH THE REGULAR PHYSICAL EYESIGHT.

I MAKE THIS DEPOSITION OF MY OWN FREE WILL AND WITHOUT A MENTAL RESERVATION OF ANY KIND.

James O. Woods
Orrville Woods

SUBSCRIBED and sworn to before me this ____30____ day of ____April____, 19_57_.

Warren S. Taylor

My Commission Expires Oct 25, 195_ Notary Public in and for said County and State.

(NOTARIAL SEAL)

BIBLIOGRAPHY

"Man or Matter" by Ernest Lehrs, Ph. D.,
Faber & Faber, London.

"Atlantis to the Latter Days," H.C. Randall–Stevens,
Aquarian Press, London.

"Flying Saucers on the Attack," Harold T. Wilkins
Citadel Press, New York.

"Flying Saucers Uncensored," Harold T. Wilkins
Citadel Press, New York.

"Flying Saucer Conspiracy," Major Donald Keyhoe,
Henry Holt, New York.

"Flying Saucers" An analysis of the Air Force Project Blue Book, Special Report No. 14, by Dr. Leon Davidson, 64 Prospect Street, White Plains, N.Y.

"The Power Within" by Dr. Alexander Cannon,
E.P. Dutton &Co., New York.

"The Psychic Sense" by Phoebe Payne & Laurence Bendit M.D. E.P. Dutton & Co., New York.

"The Power of Karma" by Dr. Alexander Cannon.

"My Experiences Preceding 5,000 Burials"
by Hamid Bey.

The above books may be obtained from the New Age Publishing Co.

Other Books by Trevor James Constable

***THE COSMIC PULSE OF LIFE*: *The Revolutionary Biological Power Behind UFOs*.** This book presents evidence that UFOs are mainly invisible and consist of both physical craft and living, biological creatures. The author convincingly shows that our atmosphere is the home of huge, invisible living organisms that are sometimes confused with spacecraft when they became visible. Mr. Constable has photographed both types of UFOs with special infrared film, some of which are reproduced in this expanded and updated edition. Despite his bold leap into the future, the general public and official ufology have a hard time accepting the evidence. In recent years teams of engineers and technicians in both Italy and Romania, unaware of Constable's earlier discoveries, obtained virtually identical infrared photos of UFOs, which were published in Italy. In 1996, NASA used ultraviolet-sensitive videotape to record swarms of invisible UFOs that looked like Constable's earlier photos. Examples from these photos are also contained in this book. Also covered are earlier pioneers into important life energies that play a big role in this research, including Wilhelm Reich, Rudolf Steiner, and Dr. Ruth B. Drown. This is an important book, recommended for those interested in the higher realms of our physical reality. **ISBN 9781585091157 • 364 pages • 6x9 • 29.95**

***HIDDEN HISTORY, RAIN ENGINEERING AND UFO REALITY*: *The Best of Trevor James Constable*.** This tribute book covers three subject areas that span over a half-century in the life of Trevor James Constable, who passed away in April, 2016. The rain engineering section documents the evolution of the cloudbuster technology based on the work of Wilhelm Reich, along with the resulting proof of success the author experienced from around the world. If taken further, this technology could be a tremendous help to humanity. The hidden history area involves little-known military men and their amazing heroics. The chapters on reality of UFOs cover flying, biological creatures that were found to reside in the hidden ultraviolet spectrum of our skies. When witnessed, these beings can be mistaken for unknown craft that are thought to be piloted by humans or aliens rather than being alive in their own right and resident to the earth. The compelling scientific proof presented here leaves little doubt that we have much more to learn about ourselves and the world around us. **ISBN 9781585091485 • 260 pages • 6x9 • 22.95**

Available from The Book Tree 1-800-700-TREE (8733)
or amazon.com, Barnes&Noble. com, or any reputable bookstore

Subscribe to FATE Magazine Today!

- Ancient Mysteries
- Alien Abductions
- UFOs
- Atlantis
- Alternative Archaeology
- Lost Civilizations
- And more ...

FATE covers it all like no else. Published since 1948, FATE is the longest-running publication of its kind in the world, supplying its loyal readers with a broad array of true accounts of the strange and unknown for more than 63 years. FATE is a full-color, 120-page, bimonthly magazine that brings you exciting, in-depth coverage of the world's mysterious and unexplained phenomena.

1-year subscription only $27.95
Call 1-800-728-2730 or visit www.fatemag.com

Green Subscriptions. E-issues.
Only $39.95 for the entire year!
Go green. Save a tree, and save money.

12 issues of FATE delivered electronically to your computer for less than $3.95 an issue. • Receive twice as many issues as a print subscription. Includes six regular issues plus six theme issues (UFOs; Ghosts; Cryptozoology & Monsters; Nature Spirits & Spirituality; Strange Places & Sacred Sites; and Life After Death). • Free membership in FATE E-club (save $10). • Free all-access to Hilly Rose shows (save $12.95). • Members-only video interviews. • Discounts on all FATE merchandise. • Monthly Special Offers.

CPSIA information can be obtained
at www.ICGtesting.com
Printed in the USA
LVHW030814251119
638398LV00013B/299/P